INNER NAVIGATION

WHY WE GET LOST
AND HOW WE FIND OUR WAY

ERIK JONSSON

SCRIBNER

NEW YORK   LONDON   TORONTO   SYDNEY   SINGAPORE

SCRIBNER
1230 Avenue of the Americas
New York, NY 10020

SCRIBNER and design are trademarks of Macmillan Library Reference USA, Inc.,
used under license by Simon & Schuster, the publisher of this work.

For information about special discounts for bulk purchases,
please contact Simon & Schuster Special Sales: 1-800-456-6798
or business@simonandschuster.com

DESIGNED BY ERICH HOBBING

Text set in New Caledonia

Manufactured in the United States of America

10   9   8   7   6   5   4   3   2   1

Library of Congress Cataloging-in-Publication Data is available.

ISBN 0-7432-2206-7

ACKNOWLEDGMENTS

A daunting task for me,
who has little to be proud of
and much to be grateful for.
How can I do justice in words
to all the givers
for all I have received?

My gratitude goes to:

First and foremost, Professor Don Norman, at the time chairman at the Cognitive Science Department at UCSD, who read my many essays on way-finding and gave me well-deserved, but always positive, criticism combined with much needed encouragement, and who finally strongly admonished me to collect them into a book. Without him it would never have been written.

Professor Ed Hutchins at Cognitive Science, who introduced me to the mysteries of Polynesian and Micronesian navigation, my first field of interest. Our discussions led me to widen the quest to human way-finding in general.

Professors Gilles Fauconnier, chairman, and Marty Sereno at Cognitive Science, with whom I had many inspiring discussions. Their interest in what I was doing was most stimulating.

Dr. Kechen Zhang, at the Salk Institute, with whom I discussed the similarities between human and animal way-finding, and who led me to many seminal articles about experiments with animals. He read and commented on my first draft of this book and kept encouraging me strongly to get it published.

Professor Bleckmann in Germany, who told me about his highly interesting turn-around experience during a forest hike in his youth.

My Canadian friend Anne Mills, who gave me precious information about her experience walking in a circle in the forest.

My hiking companions Virginia Flagg and Louis Brecheen, and also Lorna and my daughter Irline, who told me about their interesting reversals of orientation and mis-orientations. And many others with whom I discussed this subject and who provided valuable stories from their own experiences.

The Royal Institute of Navigation for giving me permission to quote extensively from Colin Irwin's fascinating article about underwater way-finding in their *Journal of Navigation*.

The journal *Oceania* for permission to quote from David Lewis's very interesting article about Aborigine way-finding.

Jake Morrissey, senior editor at Scribner, his assistant Ethan Friedman, and Brant Rumble, who streamlined my stack of essays into a book and caught the mistakes that old brains are heir to.

INTRODUCTION

Once upon a time a strange man wandered into my office at the University of California, San Diego. Erik Jonsson was his name, and navigation was his game. "I think people have an inner compass," he told me, "and when it goes wrong, the most amazing things happen." And then for over an hour he told me tales of confusion, of mental maps and their relationship—or lack thereof—to the real maps and the real world, and of his quest for understanding through the scientific literature.

I was fascinated by the stories and today, over ten years later, I still maintain that fascination. Erik, I soon learned, had a deep understanding of the scientific literature and a masterful understanding of real human behavior, especially the abnormalities of locating oneself in space.

I encouraged Erik to write his stories, to deepen his search. This book is the result. The stories charm, delight, and inform. I have learned more about the science of human navigation through Erik's writings than through the many scientific journals that I read.

We are indeed fortunate that Erik has such a remarkable sense of direction and location that, when he becomes disoriented, he can use the experience as a significant event that reveals clues to the workings of the mind.

DON NORMAN

Professor of Cognitive Science Emeritus,
University of California, San Diego;
president, UNext Learning Systems;
author of *The Design of Everyday Things,
Things That Make Us Smart,*
and *The Invisible Computer*

BEGINNING STATEMENT

The subject of this book is science—
environmental cognitive science, to be precise—
but it is still a very human book,
for I am a very human being
with all its weaknesses,
all its failings,
all its shortcomings,
and all its greatness.

INTROITUS

Our spatial ability is a truly wonderful system.
Therefore, unless one is able to come up with a truly
wonderful description,
it is bound to fall short.
What I will do is to show it from different angles
and to hold up important details for closer scrutiny.
This should enable you, my dear, attentive reader, to form
a good picture of it in your mind.

I do not make experiments,
I analyze experiences;
a method no longer in vogue,
but still quite fruitful
when done with care.

By reading this book
you will become aware of things deeply embedded in your
        mind.
The book will bring to the surface
what you only had a vague hunch about.
You will also discover things
you never imagined existed there.

PART 1

STARTING OUT

In this part I will tell two true stories from my youth, strange happenings that intrigued me for the rest of my life, demanding an explanation.

Without the insight they provided, this book would never have been written.

CHAPTER 1

STRANGE HAPPENINGS

I will start by explaining what happened to me in 1948 when I got "turned around" in Cologne. It was quite a long time ago, but the memory is still very vivid.

I arrived in Cologne early in the morning while it was still dark after a sleepless night in an overcrowded train from Ostend, Belgium, and slept for a couple of hours on a bench in the Central Station. After daybreak I set out towards the Rhine to find a steamer for a cruise on the river. I knew I was near the river and was puzzled when it never came in sight. Finally I asked for directions and was told to turn around. There was the river, and the steamers were in plain sight. I had been walking in the wrong direction, east instead of west, as I thought. Then I saw the sun coming out of the mist above the steamers. *Sunrise in the west!*

Well, that explained it. I must have become disoriented when I arrived by train during the night, so instead of going east I had actually been walking west, away from the river (Figure 1).

Now that I knew what had happened, everything would be all right, I thought. Not so! However much I told myself that the sun had to be in the east in the morning, I still felt that it was in the west and when I came to the Rhine I "saw"

FIGURE 1, TOP. *My memory picture of Cologne as I saw it when arriving from Belgium in the west. This is when the orientation of my cognitive map of Cologne was determined.*

FIGURE 1, BOTTOM. *Since my cognitive map was reversed, I saw the sun rising in the west the next morning.*

it flowing south. There was no way my reasoning could change my inner conviction. Apparently some mysterious direction system operating below the level of awareness had decided for me once and for all that in Cologne north was in the south. No appeal was possible.

This unresolvable conflict between the erroneous gut feeling and my rational knowledge was most unpleasant, and the harder I tried to resolve it, the more I suffered. I worried how long it would last: would all of Germany be turned around?

But once the boat got under way and out of Cologne, things straightened out. It was a relief to see the sun in the east again and the river flowing north. Apparently I had not gone crazy after all.

I returned in the evening when the sun was setting in the west and saw the tall spires of the Cologne cathedral in the north as we were going downstream towards it. As we approached the moorings, the area I knew from my morning walk came into view and then the most extraordinary thing happened. In an instant the universe spun around 180°. A very strange and most unpleasant experience. Afterwards the boat was going south, the Rhine flowed in the wrong direction, and the sunset was in the east. I left Cologne on the next train![1]

Only many years later did I figure out what had caused the reversal. Coming on a train from Belgium in the west, I assumed that I was looking over towards the east bank of the Rhine (Figure 2, top). But while I was asleep the train had crossed the Rhine. Therefore I was actually looking

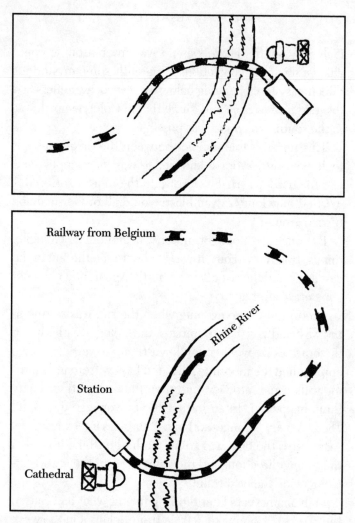

FIGURE 2, TOP. *My reversed map of Cologne.*
FIGURE 2, BOTTOM. *The correctly turned map of Cologne.*

over towards the west bank where the center of Cologne is located (Figure 2, bottom).

There is a rule among researchers engaged in experiments that "one time is no time," meaning that if something happens only once it does not prove anything; it must be repeated to gain credibility. I will therefore describe what happened to me in 1954 when I got "turned around" in Paris.

I arrived at the Gare du Nord railroad station in the morning after a long, tiring train journey from Sweden and took the métro (subway) to the Château d'Eau area, where I found a room in a hotel. As I looked out of my window I saw the old city gate called Porte Saint-Martin in what I felt was the north despite the fact that the map showed it was in the south. My sense of direction must have become confused during the underground ride, so that when I emerged into the street at the Château d'Eau station, my "automatic pilot" jumped to the wrong conclusion and made me feel that north was south. I stayed at the hotel for a week and made several attempts to straighten things out by approaching the hotel on foot from different directions, but as soon as I entered the area around the hotel I invariably had to go through the nasty 180° turn of the universe. A most humiliating experience for a Swedish forest hiker, who prided himself on his good sense of direction.

Finally I gave up and resigned myself to the fact that for me there was an "island" in Paris around the Château d'Eau métro station where my "inner compass" was turned around so that north was south and east was west. In order to avoid the rather upsetting sudden turnaround I experienced when I entered the area by street, I had to take the métro there. However, when I left the area on foot, I did not notice any sudden turnaround. It seemed more like passing through a zone of uncertainty that ended when I noticed familiar landmarks farther away.

Now my own experiences count for very little, since my honesty is not generally known, except to my acquaintances. I will therefore call a witness, Harold Gatty, who tells in his book *Nature Is Your Guide* how he got turned around at the Roosevelt airfield near New York.

> It was my first flight to Long Island; and after a long and trying nineteen hours flying, we landed at Roosevelt Field in very bad weather. It was in the middle of the night. I was overtired, and when I oriented myself on leaving the airfield I established the wrong directions. I kept feeling that I knew my directions perfectly and kept finding that I was absolutely wrong. What is more, on subsequent visits I carried my capacity for illusion with me. Everything seemed reversed. I made mistakes so consistently that I began to lose all confidence in my ability to find my way.[2]

I hope these stories have made you nod in agreement because you have also experienced these kinds of turnarounds. For I have been told many times, when talking to people about this, that something similar had happened to them. They are often a little shy about it, thinking it means that they have a bad sense of direction when in reality it means just the opposite, that they have a good one. In fact, this sort of reversal cannot happen to somebody who has a bad sense of direction, as we will see.

These three experiences point out that humans must have a "direction sense" that we are not aware of. Our minds have a directional reference frame that we rely on to orient ourselves. It is the mainstay of our spatial system. We know in which direction to go, but if we were asked how we know, we would have no answer. It is automatic. When something goes wrong so that it works against us, it becomes a terrible nuisance; when it works right, it is a tremendous asset. And it does work right most of the time—astonishingly so. I only noticed it failing twice between the ages of twenty and forty-five. One could surmise that before maturity it would not be fully developed and after middle age it would start going downhill. Now that I am past seventy, I have to admit that my sense of direction is far worse than it used to be.

It is also noteworthy that in all three cases I have recounted there was severe fatigue involved—not only general fatigue from lack of sleep, but also a special kind of fatigue from prolonged traveling that might have had an adverse effect on the spatial system, and could be labeled "directional input overload."

There is another neural system that operates at a level below awareness, the one that keeps us in balance as we stand and move around. It is automatic: we rely on it unthinkingly, taking it for granted, but when something goes wrong with it, we get in big trouble.

I vividly recall one particular midsummer night in Sweden in my youth when I happened to imbibe excessive amounts of alcohol. The result was spectacular. The horizon started seesawing in a marked manner, and to add injury to insult, it occasionally came up so that the ground hit me in the face. All the time I was firmly convinced that my body was perfectly vertical, it was just the horizon that misbehaved!

Alcohol changes the viscosity of the liquid in the semicircular canals of the ears, which throws into disarray the finely tuned vestibular system whose job it is to keep us upright and hold the picture of our environment stable. As a result, we feel we are upright when we are in fact reeling, and we see the horizon and our whole environment move.

Now that we have established that we possess a kind of inner compass, it is time to explore another aspect of our spatial system, the "mental map" or "cognitive map" we make of our environment. Our inner compass is useless without an inner map telling us our location and the direction where we want to go.

PART 2

COGNITIVE MAPS

In this part I try to describe our cognitive maps. This is not an easy task, since it is an elusive subject. When we first explore an area, we make a cognitive map of it, a map that guides us on later visits. Most of it takes place automatically in the unconscious part of our mind: we are not truly aware of what is going on. This can be confusing, but after reading this section—which looks at cognitive maps from different angles—I am convinced you will have acquired a good introduction to the subject.

CHAPTER 2

DESIGNING A WORKABLE

SPATIAL SYSTEM

Let us assume that we are in the admirable position of the Creator (for religious people) or Mother Nature (for romantic biologists) or evolution (for unromantic biologists) and have to design a workable spatial system for Homo sapiens. One thing we can be sure of: the "readout" from such a system, the information we used to find our way, had to be very simple. After all, we were not very sophisticated in the beginning: we had not even learned to make stone axes yet.

The system also has to work very fast. When we were face to face with a lion on the African savanna there was no time for long deliberations. We had to take off at once in the right direction when running for safety. Perception had to be followed by action instantly—just as we pull away the hand instantly when we touch something hot—or we would have been done for.

If our shelter were hidden from our view, for example behind a ridge, we had to be able to "see" exactly where it was behind that ridge. Either the ridge had to be transparent for our mind's eye (Figure 3, top) or we had to be able

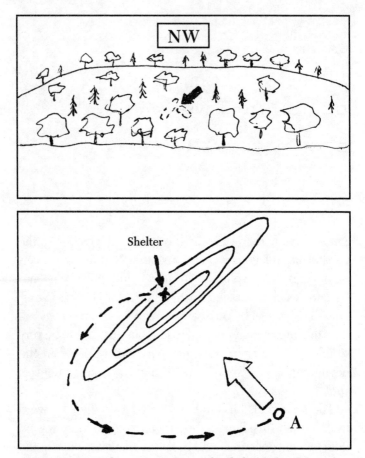

FIGURE 3, TOP. *The arrow points to the shelter behind the ridge, seen in the correct direction and location in the cognitive map.*
FIGURE 3, BOTTOM. *The hunter's dead reckoning system automatically updates his location as he walks along. He therefore "sees" the shelter in its correct direction from A.*

to mentally lift ourselves high enough to see the shelter behind the ridge.

There is another condition that had to be satisfied, given the hurry we were in: as soon as we actually saw the shelter when we had reached the top of the ridge, we had to recognize it without delay and without fail. That means that the picture we had of it in our mind—the readout from the spatial system, in other words—had to correspond as closely as possible to what we did see when it finally came into view.

That rules out everything seen from on high, a vertical view, like the topographic maps we are so familiar with (so familiar, in fact, that it is difficult for us to imagine that a map can look any other way). Instead we had to get a mental picture of the shelter seen from eye level and from the direction we came from.

Our mental picture of the shelter also had to show it in the correct location. If it was halfway down the slope on the other side of the ridge, we had to "see" it just there. This would prevent us from possibly mistaking it for an inadequate shelter higher up or farther down the slope, a mistake that, with a lion chasing us, would have been our last.

By analyzing this hypothetical situation, we have now established that the cognitive map readout contains the correct direction to properly oriented objects seen from the side. Therefore, if the spatial system is to function, it *has to* include some kind of direction monitoring system, an "inner compass" which constantly provides a reliable directional reference frame for the cognitive map.

Even if some of us have never experienced the existence of this direction frame forcefully (and even painfully when something goes wrong with it, as in the previous chapter), we would have to postulate its existence, since it is the very foundation of our spatial system. It is what we call the "sense of direction," and the fact that we intuitively label it a "sense," coequal with the five primary senses, attests to its importance.

But even if our direction frame is in perfect order, we cannot tell the direction we want to go in unless we know exactly where we are. Direction is from *here* to *there;* if the *here* is unknown, the *direction* cannot be known. Thus we need a system that monitors our movements and integrates them so that we constantly know where we are. One could call it the position monitoring system, or as navigators say, the "dead reckoning" system (Figure 3, bottom). It is Mother Nature's version of the high-tech inertial navigation system developed for nuclear submarines. The fact that in Sweden the spatial ability in general is called *lokalsinne,* which literally means "sense of location," attests to the importance of the dead reckoning part of our spatial system.[3]

This should do it: a spatial system with a cognitive map readout, which is a picture of our surroundings including not only what we can actually see but also the familiar areas hidden from our view, all seen from where we are at the moment and thus from the side. There are two necessary support systems for this readout, the dead reckoning system, which tells us where we are, and the direction frame, which tells us in which direction we are looking.

I challenge anybody to come up with a better system!

CHAPTER 3

AN INTRODUCTION

TO COGNITIVE MAPS

Navigation is knowing where you are and how to get to where you want to go. In an unfamiliar area this means that you have to use a map and a compass to find your way. But if you know the area you need no such help. You *know* where you are, and you *know* how to get to where you want to be next. It is all in your head; you have "a map in the head," a cognitive map to go by.

But when you try to look at this cognitive map, you find this is not an easy thing to do. The map is so deeply embedded in your consciousness that you normally are totally unaware of it, in spite of the fact that you use it all the time.

Part of the trouble we have when we try to look at our cognitive map comes from the "map" label, which is misleading. For our cognitive map is not a map; it does not look at all like a map. It would be better to call it our "awareness of our familiar environment."

Imagine that you are on a city street that you know well. Then you obviously know where you are. But *how* do you know that? Well, you recognize the landmarks around you. The buildings you see are familiar: a department store, a

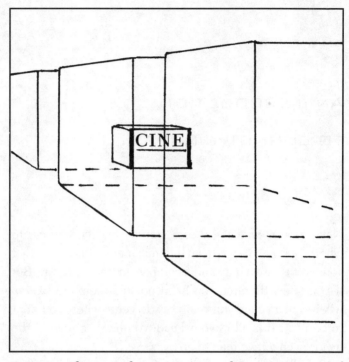

FIGURE 4. *The movie theater is "seen" in the cognitive map as if the houses were transparent.*

bank, a church, a restaurant. Those landmarks tell you where you are on your cognitive map of the city.

But in your cognitive map you can also "see around the corner." You "see" the movie theater two blocks down the street and one block to the right as if the houses hiding it from view were transparent (Figure 4). So your cognitive map does not end where your field of vision ends but covers the entire area you are familiar with. As soon as you think of a place, you become aware of the direction to it, and if you want to go there, you will "see" the route to follow, which streets to take. It is quite a remarkable system indeed, and utterly easy to use.

How user-friendly our spatial system is can be brought home to us harshly when we are in a situation where we cannot rely on it, such as when we have to drive to a place we have never been to before. Even with a good map, this is not easy. First we have to figure out where we are on the map. This can be trickier than it sounds. (I have been accosted many times by desperate motorists brandishing a street map and asking: "Can you tell me where I am?") And then we have to study the map and find the place we want to go to. We decide which streets to follow and make an effort to memorize the route. But trying to keep an eye on traffic, read street names, and check where we are on the map, all at the same time, is usually too much for us and we soon have to stop again to figure out where on earth we are.

But when we follow our cognitive map in a familiar area, we have no problem at all. We get there almost automatically. Nothing to it. Wrong, there is a lot to it, but our spatial system handles it all for us, showing us in our mind's eye our destination, when to turn, and—not least important nowadays—where we usually can find a parking space when we get there.

How can it be that our spatial system works so well for us and so unobtrusively? It is because the cognitive map is so well adjusted to the real world. In this map we see landmarks from the side, from eye level, and from where we are at the moment, which make them easy to recognize. It is as if you were comparing the actual scene with a photo taken from your position and in the direction you are looking. You could not possibly fail to recognize the landmarks.

It follows that the cognitive map, the readout from our spatial system, must change as we move around. When we turn, we "see" the map in a new direction, and when we advance, we "see" the map from a new position. This means that our spatial system must constantly keep track of our position and the direction in which we are facing in order to present us with a correct cognitive map. There must be a built-in dead reckoning system that monitors our movement, and there must be a direction frame that holds the map oriented, regardless of how we turn. When we use a topographic map we need a compass to make sure that the map is properly oriented; the cognitive map is always oriented automatically thanks to the built-in direction frame.

The cognitive map covers both the area that we see, that lies within our field of vision, and the familiar area outside our field of vision. When what we actually see corresponds to what is pictured in our cognitive map, we know where we are. The picture of the familiar area outside our field of vision tells us the direction to places we cannot actually see.

If we go towards such a place we will see it straight ahead in our cognitive map until it finally comes into view right where we expected it to be and looking very much like the picture we had of it in our mind.

We would not doubt for a moment that we had arrived at the right place, whereas when we use a topographic map and a compass to go cross-country to a place shown on the map, we have to look around very carefully indeed when we get there, to make sure that what we see actually corresponds to what is on the map. This shows not only that our spatial system with its cognitive map is a complete navigation system, telling us where we are and how to get to where we want to be next, but also that it is much more user-friendly than the awkward topographic-map-and-compass method.

But where does the cognitive map come from? The answer is that it is encoded automatically when we visit a new area. We don't tell ourselves: "Now I am in a place where I have never been so I have to see to it that I make a good mental map to use if I should come here again some time in the future." Our natural curiosity, the interest with which we look at new things, especially those that stand out as landmarks, is enough to create the cognitive map without any conscious effort.

Thanks to the direction frame and the dead reckoning system, the landmarks are encoded not only by the way they look, but also in the correct location and turned the correct way. This encoding of location and orientation of landmarks is very important, for it means that we can use several landmarks that look exactly alike as long as their surroundings, or orientation in respect to one another, are different. Let's take the blue curbside mailboxes as an example. In spite of the fact that they all look the same, they make good landmarks because they are in locations that are far enough apart to make it easy to identify them.

The cognitive map of an urban area is never complete. It covers mostly the streets that go between places we are interested in. This might seem like a drawback, but actually it is just the opposite. It is good to have a map that is tailor-made for us, showing only what we need to see. In contrast, a street map of an urban area shows mostly what we don't need, and it takes quite a bit of practice in map reading to use it efficiently, to get past the wealth of useless information and find what one actually needs.

It should also be noted that if you have lived for a long time in a place and thus have a great deal of information stored in your spatial memory, the spatial system will automatically sort out just what you need at any given moment and present you with a cognitive map showing only where you are and the streets that will take you to your destination.

Like all memories, the cognitive map of an area needs to be refreshed: it will fade if we move away somewhere else, unless we come back to visit now and then and thus revive it. The deeper the map is engraved in our memory, the better it will resist the deterioration. This ability to engrave a map seems to deteriorate, like so much else, with age. My oldest cognitive map, that of the village of my childhood encoded over sixty years ago, is still much more vivid and detailed in my mind than my maps of places where I lived only twenty years ago. If you look back at your life, you will probably find that your own childhood cognitive maps have the same astonishing clarity, even after many years.

However, when we return to a place that we have difficulty picturing clearly in our mind, we will find, to our joy and relief, that we recognize a surprising number of forgotten landmarks as soon as we see them. We can thus make a distinction between two kinds of cognitive maps:

1. the *active* cognitive map that is always available, that we can use to describe the place to somebody else; and
2. the *passive* cognitive map that also contains the landmarks that we will recognize only when we set eyes on them. Clearly the passive map is much more detailed than the active one. In fact, it is often so reliable that we can tell with certainty that a landmark we see must be a new structure since it was *not* there last time we visited the place. Quite a remarkable achievement.

COGNITIVE ROUTE MAPS

When we want to go to a faraway familiar place, we first see the direction to it in our cognitive map. We see it from where we are without any detail; it is just an awareness of a direction and a distance.

Later when we think about how to get there, we picture the route we have to follow. Usually we cannot visualize this route as one picture all the way from our starting point to our destination with sufficient detail to be useful; instead a cognitive route map is a series of pictures seen from points

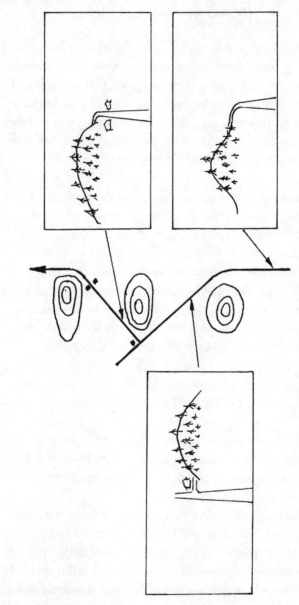

FIGURE 5. *Three views from a cognitive route map of a road.*

along the route and in the direction in which we are traveling. It is as if we, when we encoded the map of this route earlier, had a camera and snapped a picture every time something interesting appeared ahead (Figure 5).

This means that a monotonous boring part is neglected in our map, whereas a stretch with many outstanding landmarks is well covered. And since our most important task is to reach our destination without fail, all the places along the route where we could possibly take the wrong road (like forks and intersections) are shown with enough detail to prevent mistakes.

In many cases this detailed cognitive route map is not really needed. If we are on a country road, for example, in itself it leads to our destination. We simply have to stay on the road. And at crossroads, where mistakes could be made, there are signs that tell everybody who can read which way to go.

But we still "learn the road": it is an automatic process. After having driven a route a couple of times, we have a good cognitive map of the route that we follow in complete confidence. We do not even bother to look at signs that tell us what to do at intersections. In fact, road signs no longer tell us anything we do not already know, but serve, at most, as landmarks that tell us where we are along the road.

One advantage of a topographic road map, however, is that it works just as well in both directions. A cognitive route map from A to B works extremely well in that direction, but rather poorly when we go in the opposite direction. If, for example, we have driven the road from A to B many times and know it well, but never went back the same way, we will find the first time we drive on that road from B to A that it looks like a new road. The landscape and the landmarks we know so well will now be seen from another perspective and in reverse order, and will therefore be more difficult to recognize. Hence we develop *two* cognitive maps of a route to a place, one that will take us there and quite a different one that will take us back again.

But this is not all. We also need a nighttime map for a route. Even where streets are well lit, everything looks very different at night. The landmarks from our daytime map can be difficult to recognize, and many of them cannot be seen at all. Therefore when we drive a familiar route at night for the first time we are usually surprised at how difficult it is. However, our spatial system rises to the occasion and when we have driven the route a couple of times at night we develop a nighttime cognitive map of it to follow. But it is still a difficult task and we never get the same comfortable feeling of security as in daylight. After all, we are not nocturnal animals.

There are also the seasonal changes that our maps must take into account. A grove of oak trees, for example, is a solid block of green in summer and a gray patch in winter when the leaves have fallen off. And such an outstanding landmark as a lake can virtually disappear under ice and snow in winter. So we need summer maps and winter maps both for daytime and darkness. Incidentally, nighttime navigating in winter is easier than in summer because the snow reflects light very well, and so our daytime cognitive map is closer to what we actually see at night.

COGNITIVE MAPS SEEN
FROM HERE AND THERE

I am not trying to classify cognitive maps; I only want to show that there are different kinds. There are maps that show a distant place from where I currently am, "here-maps," and then those that show me a close-up of the place from a nearby point, "there-maps." For example, when I think of Julian, a charming place up in the mountains east of San Diego, I first become aware of the direction to that town. One could say that I see Julian as a little "blob" in my survey map of Southern California. Its direction I can point out, but I have only a vague idea of the distance. I cannot tell how many miles away it is as the crow flies—only a topographic map would show that; however, I know it takes me about an hour to drive there along a winding road. This is much more useful knowledge—unless you are a crow.

Then I zero in on the place and with my mind's eye I look at a set of detailed cognitive maps I have made during my many visits there, my "there-maps." I "see" the place as it looks when it first comes into view with the museum to the right and the once rowdy saloon to the left. From the main crossroad I "see" the streets heading left and right and straight ahead. I see the bakeries both from outside and inside where they sell apple pies, those sweet temptations I try (in vain) to resist. And in the liquor store I see my friend Larry, behind the counter and smiling broadly—he is a living landmark for me in Julian!

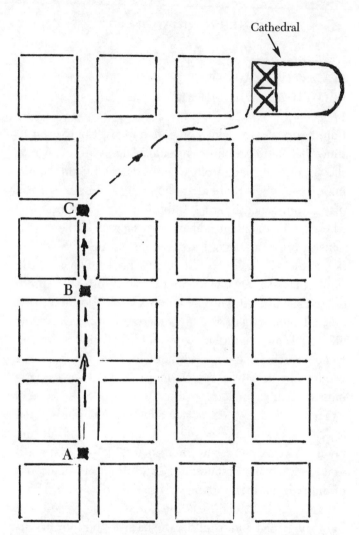

FIGURE 6. *Using a topographic map to get to the cathedral.*

All of this I see from eye level, so that I can recognize the place immediately when I get there. And my cognitive maps make it easy to recognize a postcard of the place. (But I am sure I would have trouble identifying an aerial photo of Julian; they show only roofs, and roofs are not in my cognitive maps. Who looks at roofs in a village?)

These are examples of "there-maps"—as long as I stay home, that is. But when I go there they become "here-maps," showing me how to get around and to find, like a homing pigeon, the bakery and even the shelf where my favorite pies are waiting.

COGNITIVE MAPS DERIVED
FROM TOPOGRAPHIC MAPS

As a tourist in a city new to you, you will probably use a map to find your way. Let us assume you are going to the city's cathedral. You look at the map (Figure 6) and find out that you are at point A. You will therefore have to go four blocks north and then two blocks east to get to the church. It is easy to memorize where to turn because you will pass a park on the right before you get to the cross street with the church. No need to use the map, so you put it in your pocket and start walking.

Now there are two different methods for finding the place: the obvious one, where you use the memory picture of the map, just as you would use the map itself, and a much more subtle, intuitive one, where you use a cognitive map derived from the topographic map. Normally a cognitive map is made when you explore a new area and make "pictures" of the landmarks in your memory so that you can recognize them next time you go there. But it is also possible to make a kind of cognitive map—albeit a very sketchy one—from a topographic map. In this example you will "see" the park ahead behind the houses that hide it, and also the cathedral, as if the city blocks were transparent. When you get to point B in the figure, you will "see" the park after the next cross street and you will also "see" the cathedral, but more to the right this time. When you actually get to the park at point C it comes as no surprise; even if it looks very different from what you imagined, it is unmistakably the park. With a little luck you will find a footpath that takes you diagonally through the park—a shortcut too good to pass up. As you walk along the path your mind's eye will naturally be focused on your goal, the cathedral, just a little to the right of straight ahead. Finally you reach the street and the cathedral comes into view. Instantly the rather fuzzy "preview" in your mind is replaced by the clear picture of the cathedral itself. And as you walk around outside and inside the building you develop a detailed cognitive map of the place, with what you most admire deeply "engraved" on it. This will enable you to recognize photos of the cathedral long afterwards.

How do you know if you are following the memory picture of the topographic map or the cognitive map derived from it? Well, there is a difference in these two systems.

If you are guided by your memory picture of the topographic map, you will see yourself progressing on that memory picture just as you would if you were actually looking at the map and you will consciously reason, for example: "Now I have passed the park, so I must turn right here."

If you are guided by your cognitive map, it means that your spatial system handles it all. There will be no thinking, no reasoning. You will "see" the cathedral all the time in the correct direction, and get there seemingly automatically, as if it was pulling you toward it like a magnet.

Actually it is more likely that you use a combination of the two methods. For example, you might "look" at the memory picture of your topographic map now and then so that your spatial system can transform it into a cognitive map covering only a small stretch—one block or two—ahead of you. In such a case your cognitive map will not cover the whole distance but will become only a limited extension of your field of vision.

What happens if for some reason you have to walk the same stretch again? Well, this time around, your memory picture of the topographic map will have faded—unless your spatial system is so unreliable that you have to look at the map again—and it has been replaced by the real cognitive map made automatically by the spatial system during the first walk.

One can also explore a new city without a map, just by walking "where the nose points," as we say in Sweden, and thus make a cognitive map automatically. This might seem the best way to learn about a city, but it has a drawback: Without a topographic map to give you a framework for the general layout of the city, it is easy to end up with a misoriented cognitive map—especially if one is out on a cloudy day. (My experience in Cologne is a good example.) For somebody who has a good spatial system, this would be a nuisance on a later visit, when the sun would be seen in the wrong direction. And if you ever live in the place afterwards, a misoriented cognitive map that you would be unable to get rid of would be very awkward indeed (see chapter 34, "Professor Peterson's Misery in Minneapolis").

I must mention here the tourist maps of cities that for some reason have been made with north in the wrong direction. *This should not be!* North must be towards the top of the map. Making the layout so that north is to the left or to the right or—worse still—at the bottom of the page is not a good idea. There should be a law against it, and I mean it. Instead of helping you to make an oriented cognitive map of the place, such a layout will mislead you into making a misoriented one, with north in the east, or west, or—total disaster—in the south. Sometimes the mapmakers put a little compass in the corner of the map to show where north is, thinking that it is enough. At least—if you happen to notice it—it alerts you to the danger. You can then turn the map so that north is on top, the way it should be. But then the names on the map become difficult to read, so you have to turn the misbegotten map all the time. Most frustrating. So please, *please*, makers of tourist maps, *have mercy upon us!*

CHAPTER 4

THE MENTALLY INVISIBLE

STOP SIGN

You are on your way to work. You are preoccupied. There is a problem at the job that you have to solve this morning.

At a crossroad a new stop sign has been put up. You drive past it without even slowing down. A police officer sees you and gives you a citation. You are fined, given a bad mark on your hitherto impeccable driving record, and your insurance company does not love you anymore.

You go over what happened again and again but cannot figure out why you did not see the stop sign. You are a law-abiding citizen, and a cautious, defensive driver. How could you make such a mistake?

But you have an excuse, and a very good one: You have driven that route to work for many years and know it perfectly. There is no need to *think* about how to get to work; you get there automatically, like a horse going back to its stable. After taking the same route every weekday morning, you have developed a very detailed cognitive map of it. This map contains everything you need to know about the route, including all the stop signs.

And then this new stop sign was put up, throwing a spanner in the works. You had trouble remembering it, sometimes noticing it so late that you had to stop abruptly. And so this morning, when you were preoccupied with a problem at work, you sailed right through it. How could you?

The reason is simple: This new stop sign is not yet engraved in your cognitive map. The map you made of this route over the years did *not* have a stop sign at that crossroad. And this fateful morning, when your mind had other problems to concentrate on, you followed your cognitive map, not what your eyes saw. They would just cast a glance here and there to warn about other cars and to verify position along the route. Your eyes did see the stop sign, but your brain was busy solving the problem and did not process the information.

If some joker had removed an old stop sign along the road, you would still have stopped there, since your cognitive map had it clearly marked.

To err is human, to forgive divine. But traffic courts are neither forgiving nor divine. And they know nothing about cognitive maps. If they did, they would realize that sometimes they demand that we ordinary mortals be superhuman.

Could something have been done to help you avoid such a mistake? Certainly. All it takes is to make the new sign look different from a normal sign for a couple of months until the regular route users have engraved it in their cognitive maps. How about mounting a circular white plate larger than the sign behind it? Easy to attach to all the new signs, and easy to remove later on. This would help the new map to penetrate to deeper layers of consciousness, but it will take months before the last vestiges of the old one are wiped out completely. It would be seen as a simple act of courtesy to the drivers who know the road well, and would also add to their safety.

The mistake we make when something has changed along a familiar route and we still go by the old version of our cognitive map is interesting also from another point of view. We must be by nature conservative and reluctant to change our ways. Our spatial system—that eagerly makes a *new* map of an area we explore for the first time—balks at changing an *old* one. This is good most of the time for it gives continuity and stability to our picture of our environment; after all we are designed to function in a *natural* environment where changes are relatively rare.

What happens when there is a disturbance, when we are upset about something? A fairly slight disturbance soon after the change will obliterate the new map so that we fall back into the old one and fail to notice the stop sign (Figure 7, top). Later when the new map has had time to take hold, the slight disturbance has no effect and we stop at the sign (Figure 7, middle). However, a major disturbance that shakes us will wipe out even the well-established new map and we fail to stop at the sign—much to our surprise and dismay (Figure 7, bottom).

This was demonstrated to me long ago as I walked home after a harrowing day at work, my mind still working frantically on the problem I had to solve. Coming to the front door of my apartment building, I stopped and took the key out to open the door. I was surprised to realize it was already open. And then I woke up: This door was always open during the day.

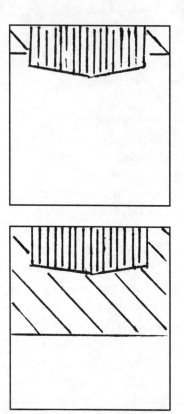

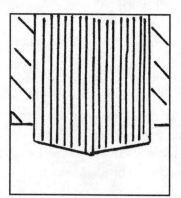

FIGURE 7, TOP. *When the new cognitive map (diagonal stripes) is recent, a slight disturbance (vertical stripes) reaches down into the old map (white).*

FIGURE 7, MIDDLE. *The new map has penetrated so deeply that a slight disturbance does not reach the old map.*

FIGURE 7, BOTTOM. *A major disturbance reaches the old map.*

It was where I had lived a couple of years earlier that I always had to open the front door. My automatic pilot, which alone had taken me home that day, had slipped back to my old going-home-from-work map, where getting to the front door meant getting out the key to open it.

Somebody might argue that this has nothing to do with cognitive maps; that this is just habitual behavior, and, of course, it can be seen that way. But I think it makes more sense to see the perfecting of our habitual behavior when we repeatedly move from A to B as the making of a detailed cognitive map of the route from A to B.

This detailed cognitive map then becomes the director that tells us what to do, that determines our behavior along the road: "Turn left at the crossroad; slow down in this curve." And: "You are now at the front door of your house. It is always kept locked. Get out your key and unlock it and go in." Or the other alternative: "This door is never locked during the day. Go straight in."

We live under the illusion that all our actions are the result of conscious observation and deduction, when in fact most of what we do in everyday life is preprogrammed, leaving our higher brain free to indulge in daydreaming, theoretical speculations, and problem solving. It's a built-in guardian angel that keeps even the most absentminded professors out of trouble.

I once heard a little story, supposed to be true, taken from an insurance claim. And I believe it, for it is simply too good to be invented. A motorist declared that, when driving home at night, he mistook the neighbor's driveway for his own and ran into a tree that wasn't there.

What happened was that the driver's spatial system "slipped" so that his cognitive map was displaced. Because of this displacement, the man felt that he was driving into his own driveway when he was in fact in his neighbor's. He therefore ran into a tree that wasn't there—in his cognitive map, that is. This is another piece of evidence that we do not function in the real world when we move around in a familiar area. We function in the cognitive map we have made of it.

After the accident, when the man fills out the insurance claim form, he looks at his cognitive map when describing what happened, and when one keeps that in mind, his explanation becomes perfectly clear and logical. The key lies in the interpretation of the word "there." To him it did not mean "in his neighbor's driveway" (as it does to the reader), or "in his own driveway" (for he was not in his own driveway). It meant "in his cognitive map of his own driveway." Because *that* is where he was when the accident happened.

In our thinking we function in our cognitive map, and not in the real world. In fact, when we move around in a familiar environment, it is in the cognitive map that we move, and what we see around us—like the landmarks we recognize—is just a *confirmation* that our spatial system is monitoring our position in the cognitive map correctly.

This explains why primitive people ("pristine" would be a more politically correct label) can find their way in areas devoid of landmarks. They trust their spatial system so well that they do not need the confirmation of landmarks.

CHAPTER 5

WHO TURNED

THE MADELEINE AROUND?

A Mr. Robb made an interesting observation concerning landmarks during a walk in Paris, as recounted by his friend Binet, a French scientist and student of our spatial system.

> "One day I was going toward the Rue Royale from the side of the Rue de Provence. At the Madeleine [a church in Paris], instead of going to the left as I should have done, I turned to the right. Reaching the Rue Tronchet, I did not recognize either the street or the gate of the Madeleine. Two or three minutes passed before I discovered that I had taken the wrong direction."[4]

It is not possible from the description to determine exactly which streets he took but it seems likely that he reached the Madeleine through the Boulevard de la Madeleine (from the east). When he got there his direction frame was reversed, so he felt he was west of the church. He therefore turned right to get to Rue Royale, which runs south from the church. In reality he was east of the church, so when he turned right he arrived in Rue Tronchet, which

FIGURE 8, TOP. *How Robb actually walked.*
FIGURE 8, BOTTOM. *What he felt he was doing.*

runs north from the church. When he got there he recognized neither the street—because he was in Rue Tronchet, not in Rue Royale—nor the portal of the church, because he was looking at the church from the north and the entrance is on the south side (Figure 8).

But why did he not see when he got to the Madeleine that he was on the east side, and not the west? The reason is that it is built like a Greek temple so that it is difficult to determine, when looking at it from the side, at which end the entrance is.

He still knew it had to be the Madeleine, for there is only one Madeleine in Paris. No other building there looks like it. He could very well have failed to recognize Rue Tronchet, however, because of his reversed direction frame, if he was not too familiar with it.

This story shows that for somebody with good spatial ability, a landmark has an extra dimension, its orientation. Robb did not recognize the north side of the church because it was facing south in his reversed cognitive map. Had it been a less familiar and/or less distinctive landmark, he might well have failed to recognize it altogether.

CHAPTER 6

FINDING CARS

IN PARKING LOTS

If we live in an urban area we really do not need much of a spatial system. There are street maps and there are signs. As long as we can read, we can find our way around. But there is one situation in which we city dwellers need our spatial system—when we have to find our car in a large parking lot.

Since this is something nearly all of us can relate to, it is a good example to consider, seeing how the spatial system handles it. For it is not always easy. Some of these lots, for example those at big shopping centers, are very large, with hundreds or thousands of cars, and it would take a prohibitively long time to look at them all.

Still, most of us manage very well most of the time. We just go to where we feel our car is. How do we do that? Well, our spatial system must have made a cognitive map of the parking lot when we drove in, parked the car, and walked to the store. In this map, our car is obviously the most interesting and important object. Therefore, when we come back to the parking lot and look out over the sea of cars, in spite of the fact that we cannot actually see our car, our cognitive map will tell us where it is. If we have excellent spatial ability, we can visualize the way the car is turned and approximate where it actually is. If we have more average spatial ability we just have a feeling it must be in some rather vague direction.

As we move through the maze towards our car, our cognitive map changes so that the car is always seen from where we are, thus we will see it coming nearer and nearer in our map. When we move sideways to avoid an obstacle, the cognitive map will take that into account and still show our car in the correct direction.

You might have trouble following this description because the use of a cognitive map is automatic, not something you are aware of. You just go to your car like a "homing pigeon" and take it for granted that you will find it where you imagine it to be. Only when the spatial system has slipped so that you fail to find your car do you realize that this is not a simple undertaking.

I remember once I "lost" my car in a parking lot and looked for it more and more desperately until I remembered I had taken my wife's car that day. I was in the right place, certainly, but searching for the wrong car.

Discussing this with a friend, he told me how, after shopping in a big department store, he failed to find his car where he was sure he had parked it. He finally discovered why. The store had two similar parking lots on opposite sides of the building, and he had been looking in the "right" place in the wrong lot.

One problem with cognitive maps of parking lots is that cars leave and new ones come and take their place. For example, you park beside a big red van, an excellent landmark that your spatial system will "mark" very clearly on your parking lot map, and then when you come back it is no longer there and you will wonder why you have trouble recognizing the place. Or you arrive when there are very few cars in the lot, so your spatial system does not bother to "mark" the location of your car very well, since you can see it anyhow, and when you come back to pick it up the lot is full and you have a hard time finding it. For our brain is opportunistic; if it thinks it can get away with a shortcut, it takes it.

I recall (with some glee) once having had to drive fifty miles to rescue my dear wife, whose car had been stolen from a parking lot—next to a courthouse, of all places. Upon arrival my rather embarrassed wife had to admit to me that the car had been found by the sheriff to whom she had reported the crime—at the very place where she had parked it. But she had an excuse. The lot has a confusing layout, with a kind of "hidden corner." That is where the good sheriff always goes to find the "stolen" cars and enjoys the mixture of relief and embarrassment on the faces of the car owners. It makes his day.

CHAPTER 7

A BACKWOODSMAN

GOES TO TOWN

Researchers studying cognitive maps do their experiments mostly in urban areas. The consensus seems to be that when people move to a new city they first make route maps and much later are able to integrate those route maps into a survey map of the part of the city with which they have become familiar. This might be true for the average city dweller, and these are the subjects that have been used in the experiments. No attempt is ever made to select those with good spatial ability or even to exclude those less well endowed.

With this in mind, I think it is high time to take a look at what happens when somebody with top-notch spatial ability, a backwoodsman, arrives in a city new to him. To make the discussion easier we choose a city that is laid out on a grid with the streets running in the cardinal directions.

In our example we assume that our man, when he arrives at the railway station, is informed that, to find his hotel, he must go five blocks east and then turn right and go two blocks south. As he walks along he makes a route map of the street going east and later of the one going south. One

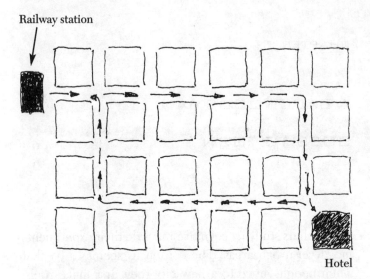

Railway station

Hotel

FIGURE 9. *The route followed by the backwoodsman from the railway station to the hotel and back from the hotel to the railway station.*

can see those route maps as a series of pictures that would make it easy for him to walk the same route again without counting the streets. One point that would be pictured especially well—to avoid overshooting it the next time around—is the corner where he has to turn right to get to the hotel.

Let us now turn to the tricky part, the part that can only be understood when one has a clear picture of how a good spatial system works. As our backwoodsman walks along, his spatial system is busy gathering material for making a cognitive map of the place. Landmarks are taken note of, like the church, the park, and, on the corner where he turns, the bank built of large blocks of rough stone to give a solid appearance, and finally the big department store filled with an astonishing amount of merchandise (Figure 9). Finally he reaches his hotel. Already he has the beginning of a cognitive survey map of the city. Let us take a look at it when he is at the hotel and turns in the direction of the railway station, the most important item on his cognitive map since he has to get back to it the next morning (Figure 10, top). He can thus "see" not only the street to the north, on which he just came down, and the street running west, but also the railway station and the street running east from it with all the landmarks his spatial system just took note of. The city blocks are transparent in his cognitive map, just as the forest is transparent to him when he finds his way in the wilderness.

There are two different processes going on simultaneously in the mind of our backwoodsman. First the obvious one, the encoding of the cognitive route map. This one we

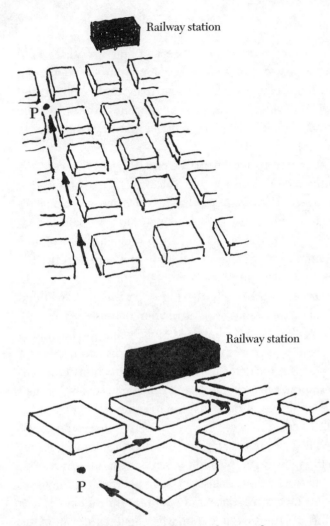

FIGURE 10, TOP. *Instead of following the route he came on to the hotel, when the backwoodsman returns to the station he chooses the street that goes more in the direction to the railway station.*
FIGURE 10, BOTTOM. *At point P the backwoodsman turns right and follows the street that goes more in the direction to the station.*

are aware of, for we might tell ourselves: "Here is the Roxy cinema. I will remember that, when I see it, I have to turn right at the next street corner." This process is present even in persons with average or less than average spatial ability.

Then comes the second part: the making of the cognitive survey map, which takes place beyond our control and below our awareness. We cannot switch the process on or off at will. It is automatic.

It is interesting to see how our man gets back to the station the next morning. Did I hear anybody say: "He must retrace his steps, go two blocks north and then five blocks west"? Maybe that is what somebody with average modern spatial ability would do. But our backwoodsman is above that. As usual, when he has to go somewhere, he follows his cognitive map, which shows him the best route. He "sees" the station approximately to the west-northwest and therefore naturally takes the street going west rather than the one going north, since he has to travel more west than north to reach his destination (Figure 10, top). This is the obvious choice and somebody with a topographic map would do exactly the same.

If our backwoodsman visited this city repeatedly, staying at the same hotel, he would every time follow the same pattern, going east from the station and west from the hotel, unless there were some overriding and compelling circumstances, like a higher concentration of attractive persons along some other route.

But he is not back at the station yet. When he reaches point P in Figure 10, top, his cognitive map looks like Figure 10, bottom, and if he is a good backwoodsman, he now turns right into the street that goes most closely in the direction of the station. By the way, next time he comes to the city and walks from the station to the same hotel he would, if he were to follow his cognitive map, turn south one block before he reached the street the hotel is on, his sensible strategy being to turn into a cross-street as soon as he feels it goes in a more favorable direction.

But why does he not go by the names of the streets rather than landmarks? Well, our backwoodsman is not used to that. There are no signposts with street names where he comes from. And he learned way-finding before he learned to read.

While this choice of streets is interesting (and somebody might want to make experiments with it to prove me right or wrong), it is peripheral to our exploration of cognitive mapmaking. What is important is that, after only one walk from the station to the hotel, somebody with a good spatial system would already have a good enough survey-type cognitive map that he could walk back to the station by a different route.

I have no proof; it all rests on my experience as a confirmed backwoodsman. There are, however, a couple of passages in articles written a century ago by two French researchers, Pierre Bonnier and Victor Cornetz, that support my view:

> We get off the train in an unknown city and follow various streets, the names of which we do not even try to remember, so that we could go back the same way. It is enough if we feel "which way" is the station we came from at every

change of direction. We have not necessarily observed the details of the route we followed, but we have always kept *the direction to our starting point* in mind, and we do not hesitate to get back to it by another route.[5]

Another quote, from Cornetz, goes into more detail:

There are people whose job it is to visit numerous cities still unknown to them, for example commercial travelers and tourists. Some of them have a remarkable orientation ability. Such a traveler . . . [walks] about town, . . . then returns to his hotel without retracing his steps. He finds it . . . by a different route from the one he came. . . . His vision has served to avoid people and vehicles . . . to recognize the hotel and the surrounding buildings seen on leaving, but it could not have served to give him the general direction of return . . . It is not his eyesight showing him the area where the hotel is located that helps him maintain the general direction during his return by unknown streets.

In this example, the role of the topographer is replaced by the memory of the route, the pedometer by a rough estimate of the distances, and the compass by the feel for the turns made "to the right" and "to the left"; it is a measurement and estimation of the angular movements in space, it is the more or less conscious sense of the displacements of the body.[6]

Cornetz, a topographer by profession, is well aware of the kind of mapping needed for the commercial traveler to return to his hotel successfully after walking around in a town new to him. And Cornetz, who had done a good deal of traveling in the Algerian and Tunisian desert with native guides, also knows that it can all be done "unconsciously, instinctively" by those who have good spatial ability.

You might object: "I could never do that!" which might very well be true. It is certainly not for everybody nowadays when most of us grow up in cities where a regular street pattern helps us in our way-finding, training grounds much too undemanding for our spatial systems to develop anywhere near to their full potential. Cornetz himself confessed that, when in a new city, he could not find his way back to the hotel without a map. He, the well-educated French topographer, goes on to compare himself with his illitcrate Adari guide, who had trouble even reciting his prayers, but whose expertise in orientation commanded Cornetz's astonished admiration. When he asked the guide at camp in the evening to point to the camp they had started from in the morning, the man "would stretch out his arm without hesitation, without thinking, without looking at the stars, and he was never mistaken."[7]

But Cornetz has some consolation, in knowing that there are those worse off than he. They are unable to point correctly from home to a place in their own city.

CHAPTER 8

DISTANCE ESTIMATES

IN COGNITIVE MAPS

One real difference between the information we glean from a topographic map and that from a cognitive map is that the latter does not give us the distances, only the directions. There is a simple reason for that: To tell the direction we only need to point. To determine the distance is more complicated, requiring a scale, and scales are a recent cultural invention. We have only been able to measure distances for ten thousand years at the most, but long before that we were able to indicate directions. The following citation, in which Victor Cornetz describes the spatial ability of his Saharan hunter-guide, illustrates this:

We have here a man who is capable of making long marches finding the shortest route to an invisible place when the horizon around him is limited by the top of grass bunches to a radius of 200 meters. He finds the shortest way not only in his large ancestral plain without landmarks but also in other plains or Saharan regions he is unfamiliar with. This man, so remarkable as far as direction and keeping on course is concerned, surprises when one asks him to estimate distances covered. He generally estimates [distance] very poorly. He will say that the distance between two points is longer than a half-day's caravan march or shorter. That is his smallest unit of measure. My hunter-guide . . . wanted to learn what an hour was, but he gave up for he was unable to make himself an unvarying representation of it. Two distances that are in fact equal seem very unequal to him if one of the marches must be made in more difficult terrain than the other.[8]

Thus the only method available to tell distance for these people is in terms of the time they take to cover it. The natural unit is then half a day's march, which is as far as they can go and come back again the same day—a rather coarse measure but a very practical one. It makes a great difference for their preparation if they know they will be able to come back the same day, compared to spending the night along the way.

For us, it makes sense to express how far we have advanced on a march from point A to point B in terms of kilometers (or miles) covered, but this is not natural; it is something we figure out by studying the map or—very approximately—by multiplying the time elapsed by our estimated hiking speed.

I am reminded here of the simple map sketches of the network of mountain hiking trails between tourist huts that are published by the Norwegian Tourist Association. They are just straight lines between the huts where trails exist, showing the hours of hiking needed with "normally fast walk without rest." This means that steep and rough trails will seem much "longer" than they actually are. Which is as it should be. This way, knowing their limitations, elderly or slow hikers can plan a nonstrenuous vacation, and young, fast ones a really challenging adventure. And since the trails are all well marked, with signposts at every junction, this little slip of paper is all the tourists need.

We can point to our destination in a cognitive map with good precision but we only have a rough notion of the distance. The reason for that is *not* a deficiency in our spatial system, but that we need to know the direction to go in from the start, whereas we need to know that we have covered the full distance only when we get there. If our goal is not actually visible at that point, we know it has to be close and can start looking for it with a good chance of finding it.

We must distinguish between the conscious rough estimate of distance available at the starting point and the much more precise unconscious feeling of having covered the full distance at the end point.

It is as so often in way-finding: the conscious versus the unconscious, the "cerebral" versus the "intuitive," and as usual the more civilized we are, the less we are able to—and dare to—rely on the intuitive part.

CHAPTER 9

PICTURES

FROM AN EXPEDITION

As we all know, vacations are much needed to compensate for a life of "voluntary slavery." (My quotes show that it is not really voluntary, nor really slavery.) But having said that, I still think it has a fitting ring to it. For some, especially those with an intellectually demanding occupation, vacation means rest and relaxation. For others, especially those who have an intellectually boring job, vacation means just the opposite, exploration and excitement; travel to new countries, new people, new languages, new customs, new food—all in order to stack up enough variety, enough excitement, enough impressions to last another year.

But memory is fickle and often fades, the more so when overloaded with an avalanche of new impressions. Memory needs help, a piece of reality to trigger the remembrance of the whole. So when tourism took off about a century ago, so did the souvenir trade. Memorabilia. Objects to buy, neither heavy nor bulky, often marked with the name of the place and—if they were good ones—recalling not only the place itself but also its specialty, like a piece of lace from Bruges, or glass from Venice.

Then came cameras—unwieldy contraptions at first, for strong backs or donkeys to carry, but practical enough for taking postcards. Tourists turned away from the trinkets in the souvenir stands and started to buy postcards instead that showed in great detail the various landmarks of the place and views from some high vantage point. Some cards even captured the life of the people: bathers on a beach, the bustle of the marketplace.

But all this was seen by other eyes than those of the individual tourist, and thus often reflected other interests, other tastes. So when cameras came down in size and price and complexity, the tourist got one of his own. He was then liberated from the frustrating tutelage of the postcard photographer and could take the pictures he liked. If he was conscientious—and some were, at least in the beginning when film was expensive—a photographer would carry a notebook and indicate for each picture when and where it was taken. But that soon became tiresome, a detriment to spontaneity. Snapshots became the order of the day. The carefully planned picture with the camera on a tripod was out; instead he snapped pictures left and right, hoping that at least one in the bunch would turn out well. No time for notebooks, unless one had brought along a secretary or an extraordinarily patient and orderly spouse!

But who cared? Looking at the pictures was for later. Finally the day came—not of reckoning but of recognition. That was when the whole stack of pictures came back from the processing laboratory. A joyous day for those with an excellent spatial memory and a frustrating day for those with a mediocre one. The pictures from the expedition had to be recognized and remembered, or they would be practically useless.

Which brings us to the cognitive part of this chapter. What is it that goes on in the mind when we look at a picture from a place we have visited? In order to identify it, we look at our cognitive map of the area, roaming around in it until we find the location the picture was taken from and the direction in which it was taken. Then the picture springs to life: We recognize the details, and we add features that are not necessarily in the picture, for example a road in a valley and a village behind a hill that we know are there because they are in our cognitive map. Since we know the direction from which the picture is taken, we can "see" what lies beyond the horizon and we "see" landmarks outside the picture to the left or to the right. The photo serves as a vehicle taking our mind back to the vacation area, and, when we get there, triggers the cognitive map of the place (Figure 11).

The photo fits right into our cognitive map because the map is also a representation of the area as seen from where we actually are—or in this case where we imagine ourselves to be—and in the direction we are looking.

Notice the difference when instead you see a picture of a place you have never visited. All you can see then is what is actually visible. It never comes to life, never jumps off the paper, never reminds you of the reality. In a view of a castle in a city, for example, the houses in the picture could just as well be dolls' houses and the castle nothing but an empty shell. You cannot see the splendid courtyard or the magnificent interiors, which would inevitably come to mind if you had actually visited it, and the picture ends abruptly at

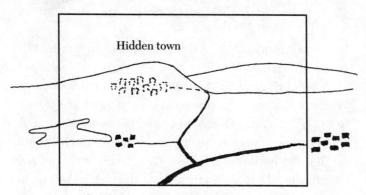

FIGURE 11. *The photo triggers the cognitive map of the area. This enables us to "see" what lies outside the frame of the photo and the town that is hidden behind the hill.*

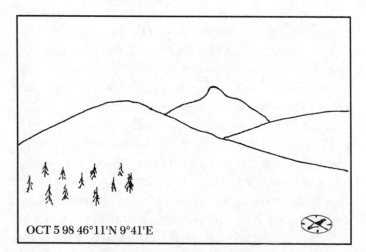

FIGURE 12. *A photo with date, location, and orientation printed at the bottom would be easy to identify.*

the edge; what lies outside is off-limits. Looking at such a picture might be pleasant, but it will never be an experience, for your spatial system cannot get involved.

It is also interesting to contemplate what happens when you look at an aerial photo of a familiar area. There is no sudden, total "gut feeling" of recognition in this case. Instead it is a struggle to figure out what the details correspond to in your picture of the place in the real world. You see, for example, a group of houses, but which one is which? Whichever way you turn it, the image never "clicks in," as an eye-level photo does. In other words, your spatial system will have nothing to do with it. Only those who are used to working with topographic maps, where the landscape is always depicted in the vertical perspective, can profit from it. This is another indication that our cognitive map does *not* look like a topographic map.

Thus a series of pictures from a vacation has a value that goes beyond what the eye actually sees. If we look at these photographs occasionally, our cognitive map of the whole region will be kept alive and stay clear. As a result, if we return to the spot, even many years later, we will be able to find our way around almost as easily as if we had been visiting there only a few months ago.

Since directions are so important in cognitive maps, it would be nice to have a camera, when on vacation, that marked on the film a tiny compass rose icon, showing in which direction the picture was taken (Figure 12). This would make recognizing the picture much easier. We already have affordable standard cameras that print the date on the film, so this ought not be prohibitively expensive.

And there are now those Global Positioning System (GPS) instruments on the market that show the latitude and longitude with astonishing precision. It would be nice, when searching for photos, to have them identified as to date, location, and direction. With today's technology, it might become reality. When I think back to the simple box camera I started out with some sixty years ago and compare it with the sophisticated photographic equipment now available, my prediction seems rather conservative.

CHAPTER 10

THE LIFE OF TRAILS

A trail is a wonderful thing. It is born and matures, it declines and disappears, just as we do. It comes to be when somebody needs to go from one location to another repeatedly.

It starts out as a dream, a desire to go to a place, say a lake for fishing or swimming. The first time it is walked it is obviously not a trail; it is just the course followed by the hiker. He sees the goal in his cognitive map and walks towards it. Even if he has to make a detour to avoid an obstacle, such as a cliff too steep to get up over, his spatial system will automatically compensate for that and still show the goal in the correct direction in his cognitive map.

The next time he walks there, he will remember some landmarks from the first hike. This will make it easier for him to find his way. Instead of having to keep on course for the whole stretch, he now has only to go in the right direction the much shorter distance from one landmark to the next.

After a few more walks to the place the number of known landmarks has increased so that the hiker never is out of sight of one. Now there is no longer any need for him to worry about keeping on course, he just walks towards the next landmark he recognizes.

Since it has not been trampled enough to show on the ground at this stage, the maker of the trail is the only one able to follow it. It is a cognitive trail, all in his head. He might, however, be able to describe it to somebody else in enough detail for this person to make his own cognitive map that's good enough to follow.

If at this stage the trailblazer stops using the trail, it disappears, except as a slowly fading memory. If instead he goes on using the trail, it finally becomes a visible trail, trampled well enough for others to follow without instructions.

It must be emphasized that as the trail develops, the earlier stages are not forgotten. The foundation for the trail is still the awareness of the direction to the lake, provided by the spatial system. And if an important landmark disappears, for example a tall dead tree falls down, this does not throw the hiker off. Its location is so well marked in his cognitive map that he will still find it. And in the final stage, when the trail is clearly visible on the ground, it still remains as a cognitive trail in the hiker's mind. If it should disappear after a snowfall, the hiker just puts on skis and follows the invisible trail guided by his cognitive map.

Even if people stop using it, the trail will remain visible for some time afterwards. In an area like the desert, where the vegetation is scanty and slow-growing, one can find trails that were abandoned centuries ago. My favorite hiking area, the Colorado Desert, is still crisscrossed by old Indian trails. They are fun to follow. On slopes they are gone, cut away by erosion, but on flat areas they still remain, looking almost as good as new in places where the soil has not been disturbed. Only the fact that shrubs grow in the middle of the trail here and there attests to its age.

The stretches where the trail has disappeared make it difficult to follow. What does one go by between the visible parts? With experience, one learns to pick the brains of the Indians who made the trail, to make one's own thinking run on the same lines as theirs.

The trick is to figure out where the trail originally went, its distant destination, and then all one has to do is to keep that in mind, and walk towards the point on the horizon the Indians must have looked at. No need to worry when the trail disappears, just go on following a few simple rules, like avoiding crests where one would be silhouetted against the sky for a potential enemy to see from afar. When one manages to do that, identifying with the mind of the trail maker, the trail will not let one down. It will reappear after each disappearance, received with the same joy and satisfaction each time. Walking like this in the invisible company of unknown men long dead, who, as it were, play hide-and-seek with you, is a most uplifting experience.

Sometimes when hiking an old trail one is surprised. It takes off suddenly to the side in the wrong direction. That is when the inexperienced Indian trail follower is tempted to improve on the trail by taking a shortcut. But one soon learns to trust the trail—to realize that those who made it knew what they were doing. When they saw fit to make a detour, it was to avoid some difficult terrain straight ahead, so one's seemingly smart shortcut always turns out longer both in time and effort.

I remember when hiking in the forests at home in Sweden. There were two kinds of trails: the old ones, which followed the dry ridges and made detours around the wet places, and the newer ones, which avoided the hills and splashed straight across the swamps.

What had happened in between? Why this difference? An innovation in footwear was the cause. Those who made the old trails wore leather shoes or even shoes made from birch bark that leaked like sieves, while those who made the new ones had rubber boots. They were modern men, those rubber-booted lumberjacks, with a job to do and no time to stop and rest on top of a hill and admire the view, restoring both body and soul. For them the walk was just lost time, a parenthesis between gainful occupations.

> *A real trail is one that has almost made itself*
> *Unplanned, with no other tools than the feet*
> *and feet make shallow impressions.*
> *So the trail has low impact,*
> *it does not impose itself on the terrain,*
> *it does not cut through hills like a raping road,*
> *it curves gently around obstacles,*
> *as if caressing the landscape.*
>
> *Sometimes it climbs hills*
> *seemingly just for the view of it,*
> *for we live not by bread alone,*
> *we also need beauty.*

A real trail, a good trail,
and all real trails are good trails,
having grown out of the human need to explore,
to get somewhere else,
such a trail has charm and is a pleasure to follow.
It fits the hiker's feet,
it puts a smile on his face and joy in his heart.

CHAPTER 11

THE ROLE OF LANDMARKS

IN COGNITIVE MAPS

Landmarks are by definition outstanding—they "stand out." They differ enough from the other terrain features around them so that we can positively identify them. They are important in way-finding since they give a fix of our position in the cognitive map of a familiar area. When we are at a landmark we know exactly where we are, and when we see two landmarks at a distance in different directions we get a cross-bearing that tells us our location.

If we mistake a landmark for another of similar shape, we lose our foothold in our cognitive map. We get lost. But for somebody with a good spatial system that cannot normally happen. He has a feeling for where he is in his cognitive map all the time so when a landmark appears he knows which one it must be just from its location and orientation. He would not mistake it for one of identical shape two kilometers away. Nor would he mistake a south-facing cliff for a nearby north-facing one as long as he has his bearings.

Thus, on the one hand our cognitive map helps us identify a landmark by giving us a feeling of approximately where it is and how it is turned. On the other hand, the cor-

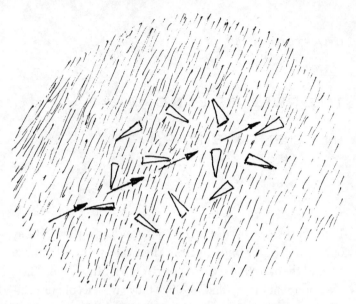

FIGURE 13. *Hiker in a dense forest can use identical but variously oriented landmarks as confirmation that he is on course.*

rectly identified landmark gives us our exact position in the cognitive map and orients it. This interaction makes it possible for a hiker to find his way in the most difficult of situations, a pathless forest where the visibility is much reduced by all the trees. There, a hiker will "pick up" landmarks on the way. In between the landmarks he will have to rely on his spatial system, which shows him the location of the next landmark. That is, he will feel in which direction he has to go to find it and approximately how far away it is.

When he arrives at a landmark, any minor error in his dead reckoning system (feeling for location) will be corrected automatically and so will any deviation of his direction frame (feeling for direction). This will make it more certain that he will find the next landmark.

It follows that somebody with an excellent spatial ability would be able to find his way in a familiar forest, even in the most difficult and unlikely case that all the landmarks (say, unusual or distinctive trees) were identical, except for orientation and of course location as shown in Figure 13. Note that the reduced visibility in the forest prevents him from seeing more than one landmark at a time. Actually, most of the time he will not see any landmark at all. Since the orientation of a landmark is well marked on his cognitive map, he cannot confuse nearby landmarks in the forest. And landmarks with the same orientation in the area are too far apart in his cognitive map for mistakes to be possible.

These examples show the importance of landmarks in our way-finding. But they are not all-important. Without our direction frame and dead reckoning system we would never find them in a terrain with low visibility!

Only in an open area where several landmarks are visible simultaneously can we manage without our "sense of direction." This had led some researchers to postulate that all our way-finding is based on landmark recognition and that a human feeling for direction does not exist. I know they are wrong, for if they were right, my ancestors in the forest of Dalecarlia would all have gotten lost and died out, and I would not be sitting here writing this.

I will end this chapter with some observations of how we encode landmarks in our cognitive map when we explore a new area. First, it is an automatic process handled by our spatial system with little input from the conscious brain. We can of course influence it by taking especially good note of the aspect of the landscape at some point, for example where we want to make sure we will not make a mistake and get lost on our way back. If we are really conscientious and worried about getting lost, we will stop and make an artificial landmark, like a little cairn of rocks at such a place.

Mother Nature did not expect us to become globe-trotters, so we are designed to stay in the same environment for our way-finding. If I get lost in the Colorado Desert, I cannot blame her: it is my own fault. She expected me to stay where I was born, in the forests of conifers up north.

I remember one time when I drove up a canyon in the Carrizo Badlands of the southern California desert. I repeatedly caught myself recognizing landmarks in spite of knowing *I had never been there before.* I could have sworn that I had already seen that side canyon, or this ridge, or that rock pile. This déjà vu was a rather humbling experience for an old wilderness hiker. I shuddered when I imagined what could happen if I actually made such a mistake when hiking in this area, where survival depends on how much strength you have left in your legs and how much water in your canteen.

The reason for this "recognition" of landmarks one has never seen before is the abundance of similar peculiar landmarks. Erosion has cut up the easily eroded rocks in that region into a maze of canyons and ridges. There are outstanding features all over the place: knife-edge ridges, rock pillars, rockfalls, narrow side canyons, sharp canyon bends, etc. But since there are dozens of each type of landmark, the problem here is to imprint a certain landmark so well in the memory that it cannot be mistaken for a similar one. I am sure the Native Americans who roamed this area long ago could do that. It was vital for their survival. They learned from childhood to notice and remember all the little details in a feature that would make it unmistakable and thus reliable as a landmark.

White men were less skillful. They were struck by the distinctive overall shape of the landmarks and failed to take note of the small details that made them unique. As a result they would "recognize" landmarks that they had never seen before and go from there thinking they would find their camp, and instead get totally lost. Hence this area got a bad reputation and name. Many a prospector went into the maze of canyons looking for gold and found death instead. Such mistakes happen not only in the wilderness. People not used to urban way-finding can mix up buildings a city dweller finds distinctive.

We have the same problem when we see people of a different race. They look so very different from what we are used to that we think they all look alike. We fail to notice the subtle differences that enable us to easily distinguish among individuals of our own race.

In some areas the problem is just the opposite. The landscape is so monotonous that the casual visitor sees no landmarks at all. But those who live there, like the Inuits on the snow-covered tundra, are adapted to these conditions and have learned to take note of subtle differences that escape outsiders.

To wrap up, for somebody with a good spatial ability, finding a landmark is not a haphazard affair. He is led to it by his spatial system; he finds it where he feels it has to be. At his level, the main role of the landmarks is not to enable him to find his way in a familiar environment, it is to *confirm* that his spatial system is working properly, to reassure him that he is on the right track.

This is very important to bear in mind. It is the key to understanding how indigenous people can find their way over large areas lacking distinctive landmarks. This will be dealt with in detail in Part 3 of this book.

CHAPTER 12

CROSSING A FIELD

In this chapter I describe the role the cognitive maps play in our way-finding. One could call the cognitive map the readout from our spatial system, showing us what we can see from where we are and—more importantly—what we cannot see!

I will take as a very simple example what happens when we cross an open field, and go on in small easy steps to the most difficult task, the crossing of a forest.

CROSSING AN OPEN FIELD

Imagine that we are crossing an open field going eastwards from our car parked on the west side of the field towards a house on the east side. There is a high mountain far away in the north and on the northern edge of the field there is a solitary tree serving as a good landmark.

Figure 14 shows how we see the mountain and the tree as we cross the field. The mountain is so far away that it stays in the north all the time, but the tree is seen in the northeast when we start out, due north when we are in the middle of the field, and northwest when we arrive at

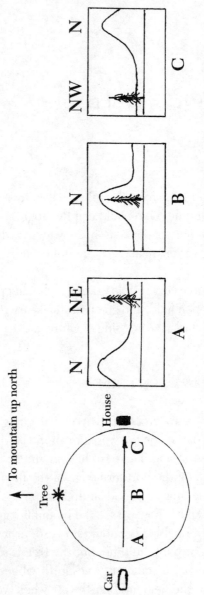

FIGURE 14. *The direction to the tree changes from northeast to northwest during the hike.*

the house. Since we are going in a straight line, the house stays straight ahead in the east all the time and the car straight behind us in the west.

As we walk along, our spatial system is continuously keeping track of what is going on. It registers the direction to the house, the car, the landmark tree and the mountain far north and makes a cognitive map out of that information. In that cognitive map we see the tree and the mountain just as we see them in reality, as shown in the figure. We also see the house in the east and the car in the west.

This is all obvious, so what is the use describing it? Well, when we look straight ahead towards the house, we actually see only the house. But, in our cognitive map we also "see" the tree and the mountain in the north and the car behind us in the west. And if we close our eyes, we will still be able to point to all these landmarks with good precision because they are recorded in our cognitive map.

CROSSING A FIELD
WITH A GROVE OF TREES

Now let us complicate matters a little by planting a grove of trees tall enough to hide what lies behind, as shown in Figure 15. We can still see the house all the time straight ahead so we have no problem keeping on course. The interesting part is what happens when we get to the point where the grove hides the tree and the mountain.

The sketch to the left shows what we see just before the landmark tree disappears from view. The middle sketch

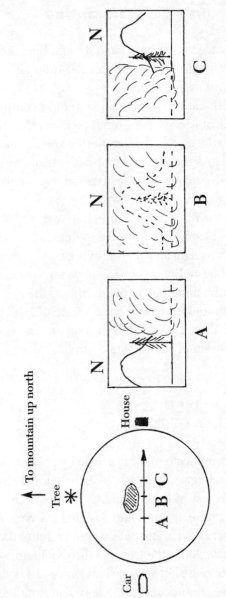

FIGURE 15. *The tree and the mountain are "seen" in the correct direction even when hidden behind the grove.*

shows how we are able to "see" the tree and the mountain in spite of the fact that the grove is blocking our view in that direction. It is as if the grove were transparent. The sketch to the right shows what we see when the mountain and the tree reappear as we have passed the grove. If our spatial system is functioning right we will actually see them just the way we expected them to look, based on the picture we had of them in our cognitive map at this point. Despite the fact that we could not see the tree behind the grove, it continued moving in our cognitive map at the correct angular speed.

This example seems rather trivial, but it is very important. It shows us how the cognitive map readout looks. When the landmark tree disappeared behind the grove we continued to "see" it just as it looked before it disappeared.

I hear protests: "This is not a map. It cannot be as simple as that! This is just a memory picture of the terrain as he saw it before the grove got in the way. In order to get a useful map of a piece of landscape, we have to see it from above."

Not so. It has to be as simple as that. When we got out from behind the grove and saw the landmark tree again, the cognitive map instantly fit the reality. The tree looked as it did in the cognitive map and we saw it in the direction indicated in the cognitive map.

And this cognitive map is much quicker and easier to use than a topographic map. If you change your mind and decide to go to the tree instead, you just walk towards where you "see" it in your cognitive map behind the trees in the grove. With a topographic map, you first have to find out where you are on the map. You then have to find the tree on

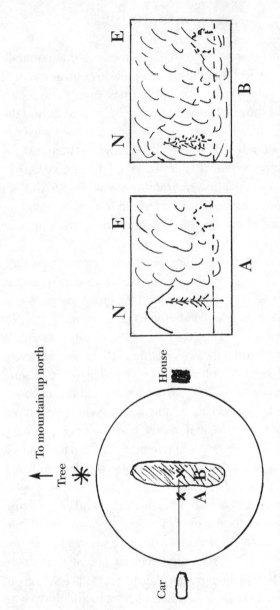

FIGURE 16. *At A the house is "seen" in the correct direction behind the grove. Even inside the grove at B the objects are still "seen" in the correct direction thanks to the oriented direction frame.*

the map and get the bearing in degrees from where you are to the tree. Finally you have to get your compass out and follow that bearing.

Imagine that the reason you suddenly changed your mind and decided to go to that tree instead is that you saw coming towards you a rhinoceros with obviously evil intentions. If you use your cognitive map, you can take off instantly and have a good chance to get up that tree before the rhino overtakes you. Using the topographic map and the compass you would be dead many times over before you got there.

CROSSING A FIELD WITH A CURTAIN OF TREES IN THE MIDDLE

Now let us put a thick curtain of trees across the field so that it actually hides the house from view (Figure 16). However, we can still walk in the right direction towards the house because we are familiar with the area and have a cognitive map of it that shows us where the house is. And the mountain in the north locks in our direction frame so there is no risk that we will deviate from our course.

Things get trickier when we go through the strip of forest in the middle of the field. The mountain up north disappears behind the trees and we have seemingly nothing to go by. However, in our cognitive map we still see the house ahead and the mountain in the north, courtesy of our spatial system. And as long as our direction frame does not start turning, it will keep us on course. And even if the direction

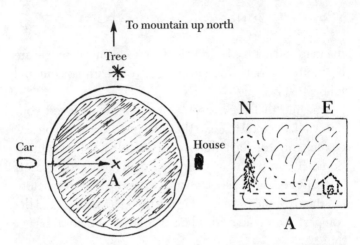

FIGURE 17. *Somebody with an excellent spatial ability still "sees" the landmarks in the correct directions after hiking to the middle of the forest.*

frame should come adrift, the distance we have to cover before we get out in the open again is so short that it would not make us deviate much. We would still see the house almost straight ahead when it appears.

CROSSING A FOREST

Finally, let us make the way-finding task really difficult by replacing the curtain of trees with a forest that covers the whole area. If we believe that we always have to go from landmark to landmark in order to find our way, then getting through the forest to the house is impossible. Admittedly this task is not for everybody, but somebody familiar with the area and equipped with a good spatial system can handle it.

As we start out, we will see the house in its direction in our cognitive map and as long as we do not lose our bearings—that is, as long as our direction frame stays in the correct position—all we need to do is to walk towards where we see the house in our cognitive map. If the sun is visible, for example, that will be sufficient to lock our direction frame in the proper position. So in spite of the fact that we see nothing but nearby trees all the time (trees that are too many and look too much alike to be of any use as landmarks), we will follow the direction to the house guided by our cognitive map (Figure 17).

Thanks to the dead reckoning system, our cognitive map will also show the changing direction to the landmark tree as we proceed. This would enable us to change our mind, for example when halfway, and go to the tree to the north outside the forest instead, and be reasonably sure to find it.

Somebody with an imperfect spatial system, which, in these modern times when we are spoiled by artifacts like maps and road signs and compasses, is most of us, will find this reasoning difficult to follow. Can this really be possible? I hope that the anecdotal evidence that will be analyzed when we study the way-finding of indigenous people will be impressive enough. For the moment, I will just cite the laconic answer of Vikars Sven Danielsson, a good friend of mine in Sweden, who replied when I asked him for tales and traditions about people who had lost their bearings and started walking in circles: *"Forest people don't get lost!"*

Our problem as civilized, intelligent beings is that we are used to evaluating our situation by observation and deduction, by looking around and drawing conclusions from what we see. If we do that in this case, all we will see is trees looking more or less alike, so we draw the inevitable logical conclusion that we are lost "in the middle of nowhere." We do not need much imagination to realize that we are in a potentially life-threatening situation and, if we have a tendency to panic, we most likely will.

But we do not function on the intellectual level when we find our way through the forest; we rely intuitively on our spatial system. We follow our cognitive map, which shows us the directions to the house we are walking towards and to the landmark tree up north. We do not *know* where we are, but we are not lost, for we *feel* where we are and which way to go to get to the house.

This is yoga, the union between the intuitive and the cerebral, the holistic way of functioning in our environment. This is the Zen of way-finding. This is why forest people don't get lost.

CHAPTER 13

WHEN THE DEAD RECKONING
SYSTEM SLIPS

There is nothing more instructive in this study of the spatial system than when something goes wrong in the student himself.

I was tired after walking around in downtown San Diego for a couple of hours when I walked from the bus stop to my home where I had lived for over twenty years. I was totally relaxed and "lost in thought" since I was on a route I had followed hundreds of times.

All of a sudden, I woke up from my daydreaming and realized that I was completely lost, for when I looked around I could not recognize any of the houses along the street. I panicked. I felt for a moment as if I had been kicked out of this earth and onto another planet. Then I regained my composure and analyzed my situation. I did of course conclude that I had nothing to worry about; all I needed to do was to continue until I came to an intersection, where I could read the names of the streets on the signpost. So I went on and when I got to the distinctive T-junction (at A in Figure 18) I recognized the place and was back on track again.

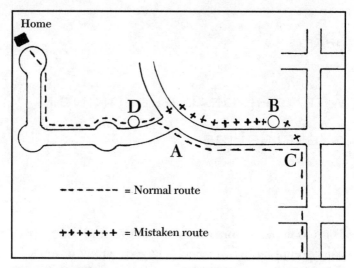

Home

D   B

A

C

- - - - - - = Normal route

+ + + + + + = Mistaken route

FIGURE 18. *When at intersection C I felt I was at the T-junction A, so I crossed over to the north side of the street to B. At B (feeling I was at D) I looked in vain for the landmarks around D and concluded I was lost.*

The key to understanding what had happened was that I found myself on the north side of the street (at B in the figure) when I realized I was lost. I must therefore have crossed over to the north side after turning left at the intersection C. I normally never cross there, since it is a detour. Where I always cross over to the north side is, instead, after turning left at the T-junction A. This shortens the walk since the street turns right, that is to say north, a little farther ahead. Apparently I must have felt I was at A when in reality I was at C, and mistakenly crossed over to the north side at C. When I woke up from my daydreaming at B, I must have felt I was at D.

But how is it I could not recognize the houses around B and therefore concluded I was lost on a street I had never seen? Well, feeling I was at D, I was looking for the houses around D. Thus my malfunctioning spatial system actually *prevented* me from recognizing the landmarks in the area. (Note that a landmark in a cognitive map has not only aspect and orientation, it also has *location*.) Therefore I might have had a vague feeling of déjà vu, but it did not pass the threshold to recognition. It was only when I got to the unmistakable T-junction that I recognized where I was.

There is another factor that can have contributed somewhat to this failure to recognize the houses. The more we use a route, the more we learn to rely on our dead reckoning system, and the less we need the landmarks (except perhaps for those that tell us when to turn). And what we do not need in our cognitive map is likely to fade and disappear.

But why this unreasonable panic when I felt I was lost? The answer is simple. Because it was caused by the sudden collapse of my spatial system when the landmarks seen did not fit the location felt. It was like a computer "crashing" when fed conflicting input it cannot handle—a stunning blow resulting in a gut feeling of panic on the deepest level of consciousness. It took quite a mental effort for me to overcome this panic and convince myself by reasoning that I had nothing to worry about.

This experience is very different from getting lost in an area where one only has a vague cognitive map, for example, driving to somebody's house whom one only visited years ago. In that case everything is tentative, like fumbling in the dark. One is never sure which road to choose at an intersection or where one is in relation to the destination, because one does not have a reliable cognitive map to go by. The reasoning becomes something like: "Well, the house cannot be in this street, for then I would have run into it by now." Hence, there can never be any conflict between landmarks seen and location felt. The "computer" does not "crash." It is all on the cerebral level. Annoyance and frustration, yes; unreasonable panic, no.

It follows that somebody with a weak spatial system can never experience this kind of deep-seated panic. Lacking a firm "sense of location," he goes mainly by the landmarks, so there is never any conflict.

I like to exaggerate a little to hammer home a point: *One with a robust spatial system recognizes landmarks because he knows where he is. One with a fragile spatial system knows where he is because he recognizes landmarks.*

Normally the sense of location (based on dead reckoning) and the recognition of landmarks along the route go hand in hand, mutually supporting each other. One with a robust sense of location knows where he is along the route all the time. The landmarks only serve—on the unconscious level most of the time—to confirm that the dead reckoning system is functioning properly. Whereas one with a weak sense of location has only a vague notion of where he is along the route unless he sees a landmark that tells him his location.

Another malfunction of my dead reckoning system happened when I was driving home from a visit to the dentist (no anesthesia involved but still a stressful experience). This time the "slippage" went from one street to a parallel one (Figure 19). Somewhere along the route I must have started feeling I was in the street to the east instead (the switch-over indicated by the broad arrow in the figure). One could imagine a period of intense concentration on something else other than the driving (like difficult problem solving) when I was totally disconnected from the environment, and afterwards, when I reconnected, I did so in the wrong street. All this did not take place on the conscious level; it was all handled—or in this case mishandled—by the unconscious spatial system.

I find it rather amazing that I went all the way to the intersection X in Figure 19 without noticing that I was going the wrong way. At X it was obvious, however. The intersection where I was supposed to turn right had traffic lights, whereas this one did not. I looked around in desperation and was relieved to read the name of the boulevard on a signpost, but the name of the cross street was completely unknown to me. This in spite of the fact that I have

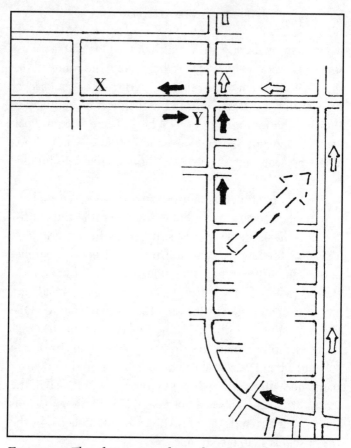

FIGURE 19. *The white arrows show where I thought I was going, the black arrows show where I actually went. The big arrow symbolizes the slippage in my cognitive map over to the eastern street. At the intersection X I realized I was lost. Seeing the liquor store at Y on my way back reconnected me to my cognitive map.*

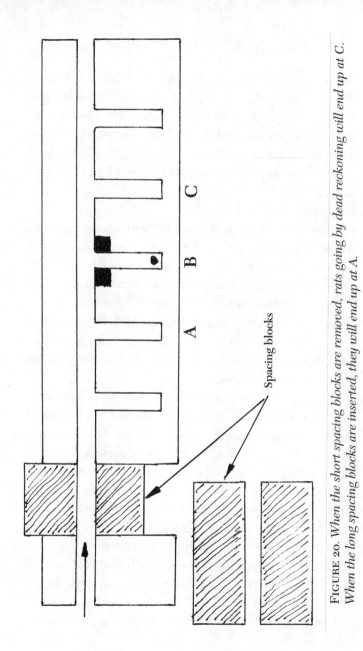

FIGURE 20. *When the short spacing blocks are removed, rats going by dead reckoning will end up at C. When the long spacing blocks are inserted, they will end up at A.*

Spacing blocks

crossed the street hundreds of times and driven along it dozens of times. But who reads street names in an area with which one is utterly familiar? However, if I had had my wits about me at that moment, I would have known where I was, for there are no other intersections without traffic lights for miles in both directions along the boulevard. But instead, I was completely confused and all I could think of doing was to continue on the boulevard hoping some unmistakable landmark would turn up. At the next intersection I still was not quite certain where I was, but at least I had enough sense to realize that I was too far west and had to make a U-turn in order to get home. It was only when I was on my way eastwards on the boulevard and saw the liquor store (Y in Figure 19) in the distance—the distinctive landmark at the intersection where I turn north towards home—that I caught up with my cognitive map, and became reconnected to my environment.

It is difficult to describe how I felt during this period, a minute or two only, from when I realized I was lost until I got back on track again: a mixed feeling—on the one hand, the comforting knowledge that I was on the familiar boulevard, on the other, a deep-seated *angst* from not knowing where I was along that boulevard, from being disconnected from my cognitive map, and thus my environment.

I am reminded here of Darwin's observation that when the sense of direction is "suddenly disarranged in very old and feeble persons," they experience a "feeling of strong distress." As my experiences show, this is true also for the "sense of location."

I must be joining the "very old and feeble." But there is

a consolation. When the spatial system malfunctions in old age we are given a deeper insight into how the spatial system normally works and are reminded of how important this system is for our well-being. Perhaps one could call such a nasty happening a "seminal geriatric spatial experience." Sounds nice and scientific, doesn't it?

The deep-seated panic I experienced when my dead reckoning system slipped so that I got lost along these very familiar routes indicates the "animal" nature of the system. It therefore ought to be easy to show in experiments that animals rely heavily on the dead reckoning system in a familiar situation, for example by lengthening or shortening the distance to a turnoff into a cul-de-sac with a food reward (Figure 20). Later I found that such experiments have already been made. And so confident were the rats in their dead reckoning that "when running in a lengthened path without cul-de-sacs turning off it, the rats sped for approximately the previous length of the path and then attempted to turn, often striking the wall."[9]

It would be interesting, however, to use the setup sketched in Figure 20 for experiments with small children and determine at what age they would start using landmarks to find the correct turnoff. My prediction is that most two-year-olds, after having learned the route to the reward properly, would go by dead reckoning afterwards, when the alley had been lengthened or shortened, like the rats, and miss the correct turnoff, while most ten-year-olds would go by the landmarks and find the reward. I also predict that some of the older children who normally follow the landmarks would regress to dead reckoning when distracted by a mental task.

Come to think of it, would higher primates, like chimpanzees, also be fooled by their dead reckoning system or would they go by the landmarks?

PART 3

DIGGING UP OLD STORIES
AND ANALYZING THEM

In this part of the book I will cite old references from the literature describing feats of way-finding that the authors found extraordinary and often impossible to explain. Most of the way-finders in the stories were the experts in this field, native people who had never seen a map or a compass but still were able to navigate as well as their civilized observers, aided by instruments, could do.

It is what is called anecdotal evidence, which is looked down upon by science as unreliable—which is true, unless it is interpreted and analyzed by somebody who himself has experienced something similar. *And—since we do not have reports of experiments using subjects with the same proficiency as these native guides—these stories are the only ones in which we can really test our theory of how our spatial system functions and see if it is able to explain the unexplainable.*

There will of course still be many old stories that I have not covered. When finding one of those, the interested reader will have the opportunity—and excitement—to apply the theory and see if it works, if it is able to give a reasonable explanation to the strange happenings.

CHAPTER 14

ADARI WAY-FINDING

IN THE SAHARA

I will start by introducing a couple of stories from the Saharan desert. But first I have to introduce the story-teller, Victor Cornetz. He was a French civil engineer working as a topographer in Algeria and Tunisia a century ago. He took time out to study the sense of direction in ants and men. As human subjects he had the hunters of the Adari tribe, employed by him as guides. They were living on the fringe of the Sahara desert in a most demanding area for way-finding.

Here is a rare combination of subjects with the highest spatial ability and an ethnographer-topographer with the education and equipment to properly evaluate their performance. However, he himself did not have a good spatial ability, which he candidly admitted. When he visited an apartment for the first time, his host had to show him the way out afterwards.

They were a peculiar combination: Cornetz, the well-educated intellectual, a product of the *"formation française,"* who always got lost unless he had a map and compass, and his Adari guides with no formal education whatsoever, who never got lost.

"Toi, tu ne sais pas!" (*You* don't know!) the Adari guide told Cornetz when he tried to keep on course unaided by his compass and failed miserably. The Adari was a kind man so he did not say it, but Cornetz understood that he was thinking: "Moi, je sais!" (*I* know!)

I really like Victor Cornetz—that charming man—that old friend I have never met: so warmly human in his writing, candidly admitting his shortcomings, gently bragging a little about his fortes, coming up with ingenious ideas, both about the subject itself and within related fields. Like his wise recommendation to students regarding the way-finding of ants (his own speciality): not to read his book first but start by studying the ants on their own. What author in a scientific discipline nowadays would tell his readers to go out and do fieldwork first, free to form their own ideas without being influenced by the author's?

But now to his first story:

From the land of Oued Souf [Algeria] to Nefzaoua [last Tunisian oasis north of the Sahara] stretches a long sandy plain [about 160 km. long from west to east and 30 to 40 km. wide . . .]. On that plain . . . the undulations are very rare, they are invisible at a distance anyhow, since the horizon is always very limited . . . There is no rocky outcropping, no dune, . . . no higher ground. On the ground there are . . . millions of small sandy humps 2 to 3 meters apart, . . . accumulations of sand caused by plants of the same species, . . . about 1.40 m. high. Here every non-Saharan is almost as lost as at sea if there are no traces of man or beast, which often is the case. The horizon is limited to 100 meters, even less because of the upper tufts of the plants. . . . On this plain the Saharan moves about without losing time, without searching. I once asked in a *douar* [a group of tents] for a shepherd to lead us by the shortest route to a point on the edge of the *areg* [the area with large dunes]. . . . The young man led us there in a straight line, without hesitation, covering a distance of 20 to 25 km, and only after traveling 12 to 15 km did the *areg* become visible far away. When I asked him to explain how he did it, he answered "*and bali* (I have it in my mind), it is over there*." He must have gone by some guidance . . . I could not perceive, but his walk was so direct, without halts, that I am led to believe it was an unconscious guidance, an instinctive guidance. One might say he had a map in his head, with the tiniest details.[10]

It seems unlikely that the shepherd crossing the desert with the grass hummocks could have relied on landmarks to keep on course for such a long distance. In this monotonous landscape, it would be like crossing a forest and finding the way by recognizing the trees! There are far too many of them and they look too much alike.

So how did he do it? As Cornetz puts it, the place was as devoid of landmarks as the sea. That ought to have given him the cue. The Polynesians were able to navigate in the Pacific finding tiny isolated islands at long distances. And the shepherd's statement: "It is in my mind, it is over there!" points to something inside.

To be fair, he comes very close to finding the solution.

He concluded the Adari was led by unconscious guidance and also that he must have had a set of maps of the area in his head.

So what is the explanation? Well, Cornetz had pointed out to the shepherd the direction to the place on the other side of the steppe that he wanted him to go to. The shepherd had marked that point in his cognitive map and as far as he was concerned that took care of the problem. All he needed to do was to walk towards the place where he saw it in his cognitive map—or as he put it, "in his mind." As long as his direction frame stayed in place, he had no problem. He would not be bothered by all the hummocks that he had to detour around, just as Cornetz would not have been bothered by them if he had seen a high tower on the other side of the steppe indicating where his destination was.

Thus the main problem is solved but a subsidiary one appears instead. What held the direction frame in the correct position for the first 12 to 15 kilometers of march when the dunes on the other side of the steppe were invisible? The shepherd did not know himself how he managed it and Cornetz could not find any explanation. In fact, he made experiments with himself as a subject trying to go in a straight line in areas where the natives had no problems. But he always lost his direction after a couple of minutes, much to the amusement of his hunter-guide. He concludes that it must be an unconscious process and that, therefore, the natives cannot explain it, which is true. The mistake he thinks he made in his experiment when he tried to go in a straight line was that he reflected and calculated instead of letting himself be carried along. And he is correct, for this is exactly what I experienced when I tried to follow a course in the desert (Chapter 20).

In this case, the Adari guide was in a familiar area; therefore one cannot completely exclude the possibility that he could have been helped in his way-finding by recognizing landmarks here and there along the way, even if Cornetz thought all the hummocks looked the same. However, in the following story told by Cornetz, he is certain that his guide was walking through a terrain that was totally unknown to him. The guide led the way cross-country from the Algerian Sahara to the Bir Stouin, well in the center of the Tunisian Sahara.

> He was often forced to deviate, he and the animal, and I followed them; afterwards he compensated automatically, without stopping, without ever looking at the sun through the day. We went on like that, 100 kilometers in five days. At one time it was so difficult that we swerved 16 to 18 kilometers to the north so as to take the dune waves on the slant. He compensated very well the day after. Finally we hit the well region with an error of 3 to 4 kilometers. There he was on familiar ground.[11]

We have here an excellent example of the spatial system in action. The direction frame automatically kept the native guide on course, and the dead reckoning system monitored his position continuously even when he had to make wide detours around obstacles. In his updated cognitive map he could therefore "see" the well in the correct direction all the time, which enabled him to "compensate automatically" for his deviations simply by walking towards the well where he "saw" it after he had passed the obstacle.

Still, it is remarkable that he could manage to do this with an error of about 2 miles in 60 (a deviation of only 2 degrees). An orienteer with a compass—running cross-country through a forest without more deviations than those necessary to avoid bumping into trees—is happy indeed if he is only 2 degrees off when he hits his target.

CHAPTER 15

THE COGNITIVE SUN

COMPASS

Except near the equator, the sun changes its azimuth (it moves sideways) as it rises in the sky in the morning and descends in the afternoon. It is not easy to figure out the direction to the sun as the hours go by; one needs a table that gives the sun's declination for the specific date and one has to use spherical trigonometry for the calculations.

That stymies the Adari or any native, right? Wrong! The Adari does not need any table, he does not need to do any calculations; he just knows. Not exactly to the degree, but close enough. This demands some explanation.

The Adari child runs around outside as soon as he has learned to walk. As he grows, he develops an excellent cognitive map of the area around the house. The sun and the shadows it casts on the ground become incorporated in that cognitive map in the approximately correct directions for various times of day and for various times of the year, a formidable amount of information to store in the memory. How can that be possible?

Well, it is difficult to explain how it is done, much less offer any proof, but I am sure that, after years of outdoor life—first as a child playing around the house, then as a youngster working as a shepherd tending the flocks in the desert around the village, and finally as an adult hunter roaming in search of game—the Adari has developed an intuitive feel for the direction of the sun at any time of day and any season. Note also that the change in direction as a function of time from day to day is very small, thus there is a continuous "refresher course" through the seasons. This feel for the sun's direction keeps his direction frame locked in well enough for him to be able to follow a course for long distances without any significant deviation.

As Cornetz describes it, the shepherd just walks right along, never hesitating, never stopping. He looks at the terrain ahead to find the easiest route between the hummocks and at the same time he automatically compares what he sees with a cognitive map of a similar landscape (similar but not identical, since he probably has never been just at that place before) seen at the same time of day at the same time of year and in the direction he is supposed to proceed. That cognitive map has the sun in it and the shadows of the hummocks and it also has direction built into it.

His spatial system then compares what he sees with his cognitive map and as long as the pictures correspond, the shepherd intuitively feels his is going in the right direction. Should they not quite correspond, the shepherd will feel that he has deviated and correct his direction, just as an orienteer corrects his course when he sights ahead with his compass and sees that the needle is not aligned with its mark.

The orienteer has to stop to do this, while the shepherd's spatial system takes care of it continuously and so automatically that he is not even aware of what is going on. So when Cornetz asks him how he does it and he answers, "It is in my mind, it is over there!" he does in fact describe it perfectly from his point of view.

One thing it is important to remember: The Adari hunter is no globe-trotter; he stays in his home area, the way Mother Nature has intended all of us to do. I did not. At age forty-seven I moved from latitude 60°N to 32°N—from Sweden to Southern California. That means that if I had developed a cognitive sun compass in Sweden, it would lead me astray in California. Did it? I do not know for certain. All I know is that when I started hiking around in the forest on the Laguna Mountain plateau, I had problems. I tried to use the same method I had developed in the forests in Sweden, where I relied on the compass only to get started in the right direction and then put it away as insurance in case something went wrong. But in California this method did not seem to work; I always deviated. I do not remember now in which direction, all I recall is that I found it surprising and most annoying.

If we continue with the assumption that a deviation occurred, it is interesting to see how it could have happened. One reason is the fact that in Sweden the morning sun moves much faster sideways than in California. Thus, if time is the determining factor, my spatial system would overcompensate for the sun's movement and, as a result, my sense of direction would drift counterclockwise. And if the sun's altitude has an influence, the rapidly climbing sun in California would result in even greater overcompensation

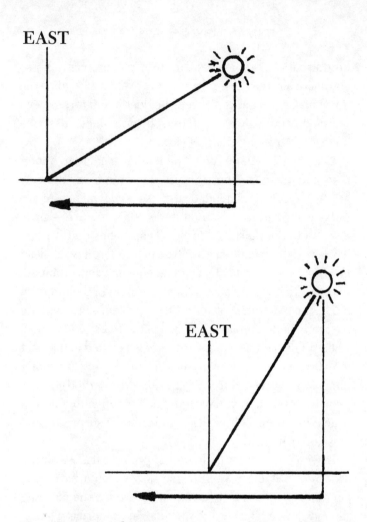

FIGURE 21, TOP. *The morning sun in Sweden rises slowly and moves fast to the right.*

FIGURE 21, BOTTOM. *The morning sun in California rises faster and moves more slowly to the right. I would therefore overcompensate for this movement and deviate to the left in California.*

(Figure 21). The deviation would thus be to the left in the morning. Using the same reasoning in the afternoon shows that even at that time the deviation would be to the left. Around noon, when the sun in California moves much faster sideways than in Sweden, there would be a large clockwise deviation but only for a short time.

The change in the movement of the sun as a function of time of day would have been really dramatic if I had moved to the southern hemisphere. There the sun moves from right to left, and thus the automatic compensation for this movement would have added to the error rather than eliminated it, resulting in a strong deviation to the left. No experiments have been done to verify this—at least not as far as I know—but people I have talked to about it, who have moved from the southern to the northern hemisphere, have reported that they had problems with way-finding in the beginning, without having any idea what the cause could have been.

I do, however, have an item of anecdotal evidence. A "northerner" on a voyage between two islands in the South Pacific felt that the ship, instead of following a straight course, veered steadily to starboard (to the right). All during his stay on the destination island he felt that the sun was rising in the south and setting in the north. The simple explanation for this strange happening is that during his voyage, with the sun as the only landmark to attach his sense of direction to, his spatial system automatically compensated for the sun's movement as if he had been in the northern hemisphere, where it moves to the right. What he felt was north was therefore moving steadily westwards. Upon arrival, what he felt was north must have moved all the way to the west, as shown by the fact that he felt the sun setting in the north during his stay there.

Note that he automatically made a cognitive map of the island—or the part of it he visited—as soon as he got there, and by the time near sunset when he noticed that he felt that the sun was in the north, which he naturally knew was the wrong direction, it was too late to correct the misorientation of his cognitive map. (Compare with my experience in Cologne, where seeing the sun in what I felt was the west in the morning did not rectify my reversed cognitive map of the place.) I know of course that anecdotal proofs, especially when based on introspection, are regarded with suspicion by scientists, but I am sure that if somebody would take the trouble to design a suitable experiment to prove me wrong, I would be proven right.

If we fly to a place in the southern hemisphere and arrive near noon, chances are that our spatial system will decide that the sun is in the south, as usual, which will reverse our sense of direction. This happened to my friend Virginia, who went to Guatemala City around midsummer. At that time and at that latitude the sun is in the north at noon. She arrived around ten A.M., but by the time she started looking around in the city, it was close to noon. As a Californian, she naturally felt that the sun had to be in the south at that time of day, which caused her direction frame to turn around. Afterwards she had a very tough time finding her way in the city since her street map did not fit her sense of direction. She tried to correct the mistake by telling herself that the United States had to be on the north side of the map, but nothing helped. Her sense of direction was too strong and not to be argued with.

CHAPTER 16

THE COGNITIVE WIND

COMPASS

Even in the Sahara the sun does not always shine. So what happens to our Adari shepherd then? Does he get lost? Well, a cloudy day in the desert is usually a windy one. Which means that he can go by the cognitive wind compass, which is easy to describe and understand.

He feels the wind on his face, feels it trying to push him over if it is strong enough and sees it shaking and bending the shrubs on the hummocks. As he starts out he knows his directions since he is in a familiar area. He can therefore integrate the wind in his direction frame, and as long as the wind does not change, he will travel in the right direction.

Thus the shepherd is going by the wind, or isn't he? If we ask him, I bet his answer would be the usual enigmatic "It is in my mind, it is over there." And he would be right! *We* would go by the wind, not he.

We would do it intellectually, determining the angle between the direction to the destination and the wind direction, and then make sure that this angle was kept reasonably constant as we walked along. That is going by the wind, and it is not easy. We would probably be quite exhausted not only physically but also mentally from all the conscious observation and calculation we had done by the time we arrived at our goal.

Our shepherd does it differently. As usual, he walks towards his destination as he sees it in his cognitive map, and this map is fixed in his direction frame which in turn is held in place by the impression the wind has on him directly (feeling the force of the wind on his face) and indirectly (seeing the shrubs bending over and the sand drifting on the ground, etc.). He makes no conscious observations, much less calculations.

Arnold van Gennep, the French ethnographer, has written a paragraph about this:

> Those who live outdoors know very well the usual directions of the wind in a certain region at various times of day. If, for example, one has had the wind against the right ear on the way out, it is enough, in order to get back, to feel it against the left ear, and one keeps oriented that way instinctively with a minimal risk of getting lost if the going and return has not taken more than 6 to 12 hours and the weather has remained unchanged.[12]

If van Gennep could successfully use this wind compass in the French Alps with the rapidly changing weather, one can assume that our Adari shepherd in the Sahara with its more stable weather pattern would manage even better.

Colonel Richard Dodge, who served on the frontier in the western United States a century ago, also noted this: "The direction of heavy winds of any season is pretty constant if not deflected by the vicinity of mountains, and it is not generally difficult to keep a course by the wind."[13]

Farther on in his book he gives a detailed account of an event in which his ability to keep on course by the wind saved his life:

> I had wounded an antelope, and was following it slowly on the broad plain, about four miles from the Buttes, when I saw coming swiftly down upon me a dense snow-cloud. I felt for my compass. I had left it in camp. Realizing the full danger of a night on that plain in a snow-storm, I at once took the only means left to me of assuring my course. If I could reach the Buttes I could find my camp. Turning my horse so that his head pointed directly to the Butte, I waited the advent of the storm. In a few moments it struck us, staggering the horses with its force, and shutting out everything beyond a circle of a few feet. Noting exactly the direction of the wind, with reference to the position of myself and horse, I started, marching with the utmost care, in a direct line, and in something over an hour was rewarded by striking the Butte.[14]

I found a similar note from the Arctic in an excellent paper by Colin Irwin, who interviewed an Inuit hunter.

> My Inuit father-in-law, Kako, is able to give names to the different principal winds that combine concepts of direction with meteorology. . . .
>
> The wind is the central and key parameter by which Kako judges direction. As he explains: "One time I was

snow blind so I had to find my way home by feeling the wind on my face. So the most important thing is the wind. When it's foggy and I can't see anything it's very hard to tell direction. But if I can get a bearing on the wind when it is clear I can then use the wind to judge direction when the fog descends. This would be in summer when there are no snow ridges, or in a boat. However I use the sun as a point of reference if the wind is shifting a lot."[15]

Continuing with the Adari hunter in the Sahara, we can ask what would happen to him if it was cloudy with no wind. Would he get lost? I don't think so. He might fall back on something I call "the fossil wind compass," a strong wind that is no longer blowing but which has left traces on the ground.

The hummocks in our example would not be round, they would have a tail of sand on the lee side of the prevailing winds (or the last sandstorm if the winds are varying). Thus every hummock would be a "sand compass," indicating the direction of the fossil wind even in perfectly calm weather. The shepherd crossing the steppe would be surrounded by sand compasses and thus be able to keep roughly on course.

I found a very good example of this fossil wind compass in Colin Irwin's paper about Inuit navigation:

> . . . on the Barrens, open sea ice or large lakes, the prevailing winds evenly etch the snow into a carpet of ridges that cover the entire hunting area. . . .
>
> The pattern is so consistent I was able to use them in

traveling from Hudson Bay to Alaska by dog team without using a compass. However, I . . . made a few navigational errors in my travels until I learned to check snow pattern directional reference against the Sun or rather an abstract conceptualization of north, south, east and west.[16]

It is clear that for the Inuits the pattern formed by the prevailing winds in the snow is a mainstay of their way-finding, which is natural since the number of clear days when they can go by the sun is very limited. But it is also evident that the sun is what they go by to determine the variations of snow pattern directions over the large areas they are covering in their hunting expeditions.

Clearly the wind can be a useful cue to directions.

CHAPTER 17

RETURNING DIRECTLY
TO THE STARTING POINT

Dr. Ouzilleau, a physician in Congo, wrote early in the twentieth century about the way-finding ability of the native hunters:

> I have had the advantage over other observers to have crisscrossed the forest with natives outside beaten trails and for no other purpose than hunting, with no other care than to find the animal that we were tracking through all kinds of terrain and forests (thickets, flood plains, marshes). I have noticed that after a day's march, having turned in all directions—which I did not keep track of—my guides always showed me the direction to the starting point. If they belonged to the region and knew the places (which made it easier for them to orient themselves thanks to the landmarks known by them), they returned directly to the starting point without hesitation. If they did not know the terrain, they still gave me the true return direction and we always came back to our starting point without ever getting lost. I dare any European to be as sure of himself as these people who never get lost.[17]

It is fairly easy to understand the prowess of the native guides who "belonged to the region and knew the places." They already had a general cognitive map of the area they were hunting in. They might not have recognized every locality they passed through during the hunt, but they still knew where they were in their cognitive map all the time, and when the hunt was over, this map would show them the direction to go in.

The hunters who did not know the terrain were in a more difficult situation. They had to make a cognitive map of the area as they went along. But if they had good spatial ability, that was no problem. The map was made automatically, and the starting point was naturally the most important part in their "instant" map and therefore the direction to it was constantly updated with good precision.

Victor Cornetz, the French topographer, had similar experiences with his Saharan guides. The setting was a steppe, a plain with no visible landmarks, covered with sand hummocks two to three meters apart, topped with tufts of tall grass that reduce the visibility to about a hundred meters. These Saharans sometimes had to go far from the camp at night to look for a camel that had escaped. After finding the fugitive they were then able to return straight to the camp. He actually made an experiment walking around at night with a guide in various directions for about an hour, whereupon he told the guide to lead the way back to the camp, something the guide did without any hesitation. When Cornetz asked him how he knew, the only answer he got was: "The camp is over there."

Cornetz was convinced that the Saharan could return

to the camp "only because he came from it a while ago." Since he, the cartographer, would need a compass, a pedometer, and a sheet of sketch paper to map his movements in order to know the direction back to the camp, he concluded that the Saharan must do the same unconsciously (Figure 22). He wrote: "There is here a kind of interior document, product of the instinct of direction. This instinct seems to be an intermediary agent between the outer environment and the visual sense to which it gives comprehensive information for the return."[18]

A very good analysis indeed. Cornetz has reached very much the same conclusion I did; he only uses different words, different labels. What he calls "interior document" is of course the cognitive map, kept oriented by "the instinct of direction," what I call the direction frame. This cognitive map acts as an intermediary to the real world, showing the direction to the starting point. For we do not interact directly with our environment; we function instead in the cognitive map we have made of it.

This was a most interesting experiment, using a subject with an outstanding way-finding ability. And even more interesting are the conclusions that Cornetz is able to draw from it. It ought to have opened the eyes of the scientists at the time. But, unable to understand what was going on in the mind of the Saharans, the researchers just swept it under the carpet as anecdotal evidence, nothing for serious scientists to deal with.

> *I really admire Victor Cornetz*
> *this remarkable man*

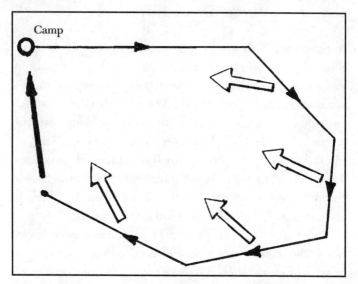

FIGURE 22. *After completing his mapping of the route, Cornetz finds the direction to the camp (black arrow). The Saharan guide is automatically aware of the direction to the camp all the time (white arrows).*

this amateur scientist
in pursuit of insight
not for money
not for fame
just for the love of it.

However, I do not agree with Dr. Ouzilleau's statement that no European would be able to do as well. I am sure people living in the large forests of Northern Europe would manage. They depend on the same spatial ability for survival and once used to the new type of terrain, they should have no problems.

Let me take an example from a northern forest with modern way-finders to show this. The Canadian researcher Romedi Passini writes:

Frederic stopped his yellow Beetle at the edge of a vast forest, stretching over an undulating landscape. Equipped with our knives and baskets, we bashed through the wood, up and down small gullies looking for mushrooms during the greater part of the afternoon. On our return, we were again surprised at the ease by which we found our point of origin, and at the precision by which we walked towards the yellow Beetle . . . as if guided by a sixth sense. (*Author's personal notes*)[19]

This "sixth sense" that guided them is of course their spatial system working unobtrusively in their unconscious minds and, when it was time to return, showing them the Beetle in their cognitive map in the correct direction.

Let me add here that in our latter days there are global positioning instruments on the market doing the same thing. One only needs to press a button when leaving the starting point and the gadget will indicate the direction back all through the hike.

CHAPTER 18

SINGING IN THE FOG

I'd now like to recount a story told by British Wing Com-
mander E. W. Anderson, one of those charming men I
have only met through their charmingly human writings:

> Spencer Chapman, the explorer, was with a party of Eski-
> mos, paddling back to their home fjord along the Green-
> land coast in their frail kayaks, when dense fog clamped
> down. To his amazement, the paddlers were unconcerned,
> laughing and splashing, but keeping a safe distance from
> the high cliffs by timing the echoes. An hour or so later,
> they suddenly all turned at right angles and almost at once
> the home fjord loomed up ahead. At every headland, a
> cock snow-bunting was staking his claim, each with a
> slightly different song. When the paddlers heard the call of
> the bird on the headland leading to their fjord, they turned
> inwards.[20]

The story follows a common pattern. The Westerner
observes the calamity with great apprehension and the
natives continue their business totally unperturbed. And
then, as a good Westerner, he analyzes what happens and
finds a sensible explanation.

The Eskimos were making a lot of noise and the sound was reflected from the cliffs. This informed them how far out they were. And then there were the birds singing on the promontories. It all sounded the same to Chapman, but the Eskimos could distinguish the individual bird arias and this enabled them to tell which promontory the song came from. Smart guys.

Thus the mystery had been solved. There was no unscientific sense of direction or sense of location involved; all the Eskimos used was keen observation and logical reasoning.

I beg to differ. The Eskimos, as always, had their cognitive map of the area to follow, so for them there was no problem: They just paddled on as before back to their village. The logical explanation was an invention by the Westerner, which saved him from having to admit that he did not understand what was going on.

This does not mean that the Eskimos did not make use of the sound cues for their navigation. I am sure they did, but I am not sure they were aware of it. However, the sounds perceived were not the basis for their navigation as the Westerner assumed, but just served to confirm that their spatial system was working properly. It was a welcome reassurance that they were on course; that their feeling for direction was correct. They would have made it even if the birds had been silent!

CHAPTER 19

A REPORT

FROM A SALTY PLACE

On February 14, 1870, Sir H. Bartle E. Frere, K.C.B., vice president of the Royal Geographic Society, read a paper about the Runn of Cutch, a featureless salt plain 150 miles long and 40 to 50 miles wide, southeast of the delta of the Indus—a desolate place: "There are no trees, no tufts of grass; and the bones of dead camels are visible for miles."[21]

During the summer months, when the monsoon blows, the Runn is inundated with salt water from the Indian Ocean mixed with rainwater brought by rivers to a depth of a couple of feet. Then the traveling is generally done by night to avoid the daytime haze and glare off the water, which can confuse even experienced guides; the stars are also usually visible.

Owing to the absence of any prominent natural features or marked tracts, the best guides seem to depend entirely on a kind of instinct—they will generally indicate the exact bearing of a distant point which is not in sight quite as accurately as a common compass would give it to one who knew the true bearing. They affect no mysterious knowledge, but are generally quite unable to give any reason for their conclusion, which seems the result of an instinct— like that of dogs and horses, and other animals—unerring, but not founded on any process of reasoning, which others can trace or follow.[22]

The Runn of Cutch is definitely a most demanding place for way-finding, a smooth featureless plain without even a noticeable slope, even when dry, a shallow lagoon during the rainy season, and nearly impossible to navigate in the day-time due to the glare. And yet the best guides know exactly the directions to locations they cannot see. Sir Bartle is right to be puzzled. The guides do not pretend to have any mysterious knowledge, but on the other hand they are not able to give any explanation for their abilities. They know but they do not know why they know. So he draws the rather natural conclusion that they must rely on an unerring instinct.

Let us try to analyze this story. During the night crossings the stars are generally visible, and we can therefore suspect that the stars play an important role in the "instinct." But it must be an indirect one, otherwise the guides would have told Sir Bartle that they used the stars to get their directions. This leads to a scenario something like this:

The experienced guide has a cognitive map of the Runn. It cannot contain anything on the surface of the Runn itself, since the Runn is devoid of landmarks. But it contains the locations of places around it. These places are in a known direction—the guide can point to them—and at a "known" distance; by that I mean that the guide has a feel for how far away they are, but might not be able to express it in words.

As he starts out into the water of the Runn, the guide knows the direction not only to his destination but also to other places around the lagoon that he is familiar with. Does he go by the stars? Yes and no. He rides in the direction in which he feels is the destination, or, more properly, where he sees it in his cognitive map. As long as his cognitive map is correctly oriented, he will go in the right direction.

How could they have learned to do this? By following experienced guides, until a single glance at the starry sky at any time of night would tell them the directions exactly, until they developed a cognitive star map which held their direction frame in the correct position. Just as we can move around in a familiar area using landmarks on the ground to keep our direction frame oriented, they did the same thing using landmarks in the night sky. Their task was more difficult because the stars move, but it was not impossible.

It should be noted that the guides did not need to have a certain star straight ahead to guide them when crossing the Runn. All they needed was a recognizable pattern of stars in *any* direction, and that would serve to lock in place the direction frame. And they did not need to know the names of any constellations, just as we use a recognizable hill on the horizon to keep oriented without knowing its name.

But we still have not answered Sir Bartle's question: How could the guides indicate the directions to invisible places when on the Runn, especially when it was under water and thus totally featureless? After all, as they ride along on the plain all these directions change, except for

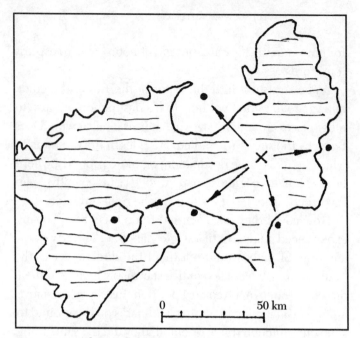

FIGURE 23. *The guide knows his position in terms of directions to places at the edge of the Runn.*

those to the starting point and the destination. They relied as usual on their cognitive map updated by the automatic dead reckoning system. At any point during the crossing they felt how far they had gone from the start and how far remained to the destination. This enabled them to point out correctly any place included in their cognitive map of the area. What seemed impossible to Sir Bartle, because all he could see in any direction was water and his cognitive map was nonexistent, was for the guides, with their excellent cognitive maps, as easy as if the places had actually been visible on the horizon.

In a way they did *not* know where they were, for they were at a featureless point "in the middle of nowhere," a useless knowledge even if they had it: but they knew the directions to—and therefore how to get to—places around the Runn (Figure 23), a far more useful knowledge.

We must make one more point, and an important one, because it explains perfectly Sir Bartle's report. Since all the calculations were taking place on the unconscious level, the guides had no idea what was going on. Even if they had the misfortune to be "civilized," they would have been in the dark. When asked to point to an invisible place they just pointed in the direction in which they knew it was. To them it was obvious that the place lay in that direction; they saw it there. It was as frustratingly obvious as if somebody asked us:

What is the shape of this ball?
It is round.
How do you know *it is round?*

CHAPTER 20

TRY TO GO STRAIGHT,

BUT DON'T TRY TOO HARD

> *It is rare indeed that a mystery disappears, one*
> *can only make it change position, but it is useful*
> *that it should change position.*
> — MAETERLINCK

Is there more to our spatial ability than meets the eye? I asked myself this question after a little experiment I made that had a very surprising result.

It started when I was hiking across a *bajada* (an evenly sloping plain) in the Colorado Desert. It was late in the afternoon and I was hiking in a southwesterly direction towards a crossroad: the easiest of targets in orienteering, virtually impossible to overshoot, since one always finds one of the roads going towards it!

On my way I had the idea to test my ability to walk in a straight line without looking at the landmark on the horizon, thus imitating the conditions of walking in a forest where the visibility is low because of all the trees. I therefore pulled down my sunhat in front so that I could only see 5 to

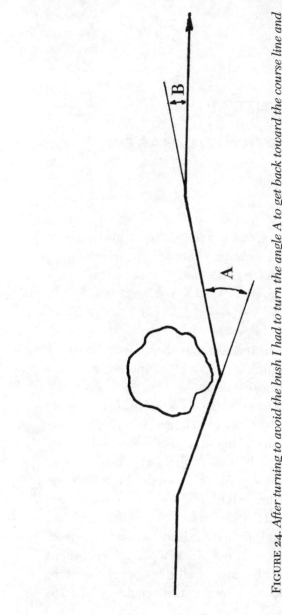

FIGURE 24. *After turning to avoid the bush I had to turn the angle A to get back toward the course line and then angle B to get back on course.*

10 meters ahead. Tired as I was after hiking many miles that day, I walked like a robot. I didn't worry about the direction, I just let my feet handle it on their own. Now and then I looked up to check how I was doing. Every time I was surprised: the landmark was straight ahead. Without really trying, without making any conscious effort, I seemed to be going towards that crossroad like a homing pigeon.

This made me wonder how I could keep going in the right direction. The *bajada* was full of shrubs and cacti, forcing me to make detours all the time. In fact, I was making more detours than I was walking straight. After each obstacle I compensated by turning to the other side once the obstacle was passed. And then when I was back on the course line I must have turned again to resume the right direction (Figure 24). If I had tried to *calculate* when to turn and how much to turn in order to get back on course, I would never have managed.

The next morning I decided to do something a little more orderly. I wanted to see how much I would deviate from my course if I walked 200 double steps (approximately 300 meters) with my sunhat pulled down so that I could only see some 5 to 10 meters ahead. I used a simple method:

I marked the starting point well, walked some 10 meters in the test direction to set a marker so that a good landmark was visible on the horizon straight above the starting point (Figure 25), and measured the bearing to that landmark.

Then I went back to the starting point and walked towards the marker with my hat pulled down so that the ground was only visible for 5 to 10 meters ahead. I counted

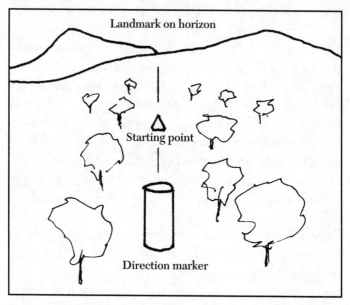

Landmark on horizon

Starting point

Direction marker

FIGURE 25. *Lining up the direction marker and the starting point with the landmark on the horizon. This landmark is used to make the return to the starting point more precise than when using a compass.*

my steps and after 200 double steps turned around and walked as straight as possible towards the landmark on the horizon. When abreast of the starting point, I measured the distance to it in double steps and calculated the deviation in degrees. I repeated this nine times, choosing courses in different directions each time. The results are shown in Table 1.

TABLE 1

Trial number	Deviation (degrees)
1	11 left
2	15 right
3	2 left
4	4 left
5	1 right
6	3 right
7	3 right
8	2 left
9	5 right

The first two trials did not turn out so well, but one advantage of being both experimenter and subject is that you know very well what is going on in the mind of the subject!

In the first trial I was all keyed up, wondering if I would be able to do it, and worrying what the result might be. I therefore made an effort to figure out intellectually, from the angle of my shadow on the ground, which direction to go in. I was very disappointed when I saw how much I had deviated to the left—11 degrees in just 300 meters.

Thus the first trial had shown that I deviated to the left, and consequently when I did the second one, I had in my mind, *"Don't go to the left now,"* repeating over and over again. I therefore deviated a whopping 15 degrees to the right.

After this humbling experience I gave up trying to do well and resigned myself to letting my feet handle it. Lo and behold, they did extremely well!

Disregarding the first two trials, the results show that I have a startling ability to walk in a straight line. I must confess I was surprised myself. The average deviation is only half a degree to the right. And the average of the deviations, regardless of direction, is 3 degrees. I would be happy to do as well with an orienteering compass! In fact the result is so surprising that I have difficulty explaining it. And in this case I constantly had to make detours to avoid running into shrubs, and especially jumping chollas, a species of cactus that every experienced desert hiker gives a wide berth to.

Of course the sun must have helped or rather my own shadow helped when the sun was behind me; and the shadows of the shrubs and cacti helped when the sun was in front, although it was never straight ahead or straight behind. And how could I have estimated the angles of the shadows on the ground with such precision, especially when walking towards the sun when there were only the shadows from the scraggly desert shrubs to go by?

The *bajada* where I was walking had a fairly even but gentle slope, only 1:50 on the average. Unless I was walking straight uphill or downhill, it is therefore difficult to see how

I could have estimated the angle against the slope direction when looking only 10 meters ahead.

Could it have been just plain luck? Once or twice, yes; three times perhaps; four times hardly; seven times, *no!*

Of course this is only a simple attempt and even though I have described it honestly to the best of my ability, it is not a valid scientific experiment since I am both the experimenter and the subject. It is, at least, a strong indication that something strange is going on that is well worth looking into more deeply. But don't expect results like this if you let loose a bunch of city-bred undergraduates in a desert or forest. Abilities like this are not learned on a sidewalk.

But how did I manage to wind my way through the cacti and stay on course without a compass? I did of course follow the direction given by my direction frame (my "compass in the head"). As I stood by the starting point looking at my marker 10 meters ahead of me, I imprinted this direction onto my reference frame. Walking along afterwards, whenever I made a detour to avoid an obstacle I felt the pull of this direction and could return to it on the other side of the obstacle as easily and correctly as if I had actually seen it marked on the ground. As long as I did not try to improve upon my performance by observing shadows and calculating angles, but let my feet do the walking, I didn't deviate.

An acquaintance, Bill Gookin at San Diego Orienteering, confirmed that an orienteer with a compass could expect to deviate 2° to 5° in 300 meters, depending on the terrain. Thus I did just as well without a compass in my little experiment. He also stated that some elite orienteers in Sweden had told him that they did not use the compass much to keep on course. Some even had a very accurate internal sense of north that they checked before they started; if their internal north was off, they might as well go home, because they would tend to be pulled off course by their skewed internal sense of direction, and doubt the compass for the rest of the day.

In the beginning of this book I described my experiences in Cologne and Paris where my "compass in the head" got turned around. Just as these mishaps demonstrated to me the astounding strength of the direction frame, this desert experience showed me its astounding precision.

Thus we come to the quote from Maeterlinck that opened this chapter.[23] The mystery is solved but in its solution another mystery appears: *What is it that keeps the direction frame in place?* As my experiment—and those of the expert Swedish orienteers—indicate, there is something going on in our spatial system that science has yet to verify.

CHAPTER 21

STRATEGIES FOR WALKING

IN A STRAIGHT LINE

In the previous chapter I described an experiment I made, or perhaps I should be more humble and call it an "experience."

It is good to have an experience like this, that shows something unexpected going on. It stirs up the "little gray cells," as Hercule Poirot calls them. Old ideas that have been sitting undisturbed for so long they have acquired the status of facts suddenly become suspect. They have to be modified, even replaced, before the puzzle can be fitted together again, which can be a painful process in the beginning. And then out of the chaos an idea is born, usually early in the morning when the brain has had a chance to work on it undisturbed for a couple of hours. Often the idea is greeted with suspicion. It has to prove itself; it has to show that, when it is adopted, a new picture emerges that makes sense. And it is an exhilarating moment, well worth the days of despair preceding it.

In the case of my desert experience, it was the realization that it is possible to proceed in a straight line without any cues to hold the directional reference frame in place that

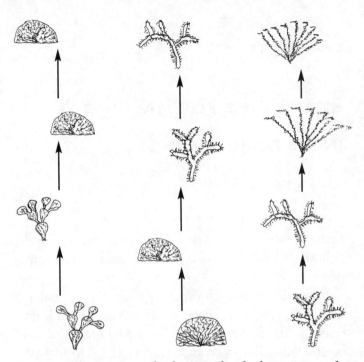

FIGURE 26, LEFT. *Four shrubs seen ahead when starting; the arrows indicate the course to follow.*

FIGURE 26, MIDDLE. *After passing two shrubs, the course line extends into the new shrubs ahead.*

FIGURE 26, RIGHT. *The shrubs seen when starting are all passed, but the course line extends into the new shrubs.*

provided the missing piece of the puzzle. It demonstrated the continuity that takes over when there is nothing to go by. For the funny thing is that even when there is seemingly nothing to hold on to, there is still something we can fall back on.

So how do we keep the direction frame in place? We can call the process the *expanding cognitive map*.

Let us assume that we are in an area where the visibility is reduced to 20 meters and where many obstacles on the course line force us to detour. Before we start out, we have looked at the surrounding area and made a cognitive map of it. This map, like all cognitive maps, is held in place by our direction frame, which means that every object in it is seen in the correct direction. The direction we plan to walk is also marked on this reference frame. The situation is shown in Figure 26, left.

After we have walked for 10 meters we will now have a part of our original cognitive map behind us and a new area visible ahead and outside of our map.

Did you notice the mistake in the last statement? The new area is *not* outside our map. Thanks to our "automatic mapping office" (a very important aspect of our spatial ability), as soon as we see new objects, they instantly become included in our cognitive map. Therefore the direction in which we are walking (seen as an imagined line in our original map) is now extended forward (Figure 26, middle).

After we walk 20 meters our new map is completely outside the original map, but the original direction is still in it (Figure 26, right).

Let us compare this situation with what would happen if we were blindfolded for a 20-meter walk away from the starting point. After removing the blindfold, we would be looking ahead into a completely new area. We would instantly make a cognitive map of it but the original direction would not (in fact, could not) automatically be part of it.

Why don't we lose our direction when we make a detour around an obstacle? It is because the direction we decided to follow is well marked on our direction frame and, as long as we are in a familiar area, this frame is held in place by all the landmarks around us. Since, as I just showed, everything in sight becomes a familiar area as we move ahead, it follows that the direction frame will remain in the correct position. Therefore when we make a detour, we will feel the "pull" of the correct direction and "automatically" return to it when the obstacle is passed.

One could even go so far as to argue that as long as we can see some distance ahead, provided our "mapping office" is in good working order and provided we are reasonably skilled overall at finding our way, we cannot get into an unfamiliar area. Wherever we go we will always be inside our cognitive map. Ahead, it will reach only as far as we can see, but behind us it will cover the area along our track all the way back to the starting point.

It is clear that this expanding cognitive map cannot stay oriented during a long march. After a kilometer or two, depending on the visibility in the area and the spatial ability of the hiker, some cues will be needed to keep the direction frame in place. When outer cues are missing, as in cloudy weather with no wind, one is tempted to postulate an inner one, most likely of magnetic origin, available to people living close to nature, who have learned to contact it, and dare to rely on it. After all, there are several species of animals that have been shown to rely on internal magnetic compasses, and who are we to set ourselves apart?

My description might give the impression that it is something we are aware of doing. Not so. It is all handled on the unconscious level. I ought to have put the word "automatically" in every sentence. I strongly suspect that the less we interfere consciously with the system, the better it will work; we should let the "animal" within us handle it, just as a rider who leaves the way-finding to the horse at the end of the day will always come back to the stable. "Brother Donkey," as Saint Francis used to call his body, knows best in this business.

I am afraid that many modern readers will find this description impossible to believe. Most of us live in an urban environment nowadays, which prevents us from developing a good spatial ability in childhood, and later in life we move too far and too fast for the cognitive "mapping office" to keep up. Even those of us who somehow managed to develop a good sense of orientation are reluctant to rely on it fully; our intellect is in the way. When it tells us that something is impossible, we believe it and do not even try. "Primitive" people do not have this inhibition. In their culture it is not an impossible thing to do, it is the only possible thing to do. We have been spoiled by topographic maps and compasses, whereas they have only "the map and the compass in the head" to go by.

CHAPTER 22

AN OLD STORY

FROM A COOL PLACE

This old story was saved from being forgotten by Charles Darwin himself, who quoted it in a paper in *Nature*. One understands why Darwin was interested, for it gives a good account of how the sledge drivers kept on course crossing the frozen sea during an expedition in 1820–1823 to the Bear Islands, north of Siberia, led by Ferdinand von Wrangell, a lieutenant in the Russian navy. Like Cornetz's reports from the Sahara, we have here the fortuitous combination of an expert way-finder in action and a skilled topographer witnessing and evaluating the way-finder's performance.

I will notice the remarkable skill with which our sledge-drivers preserved the direction of their course, whether in winding among large hummocks, or on the open unvaried field of snow, where there were no objects to direct the eye. They appeared to be guided by a kind of unerring instinct. This was especially the case with my Cossack driver, Sotnik Tatarinow, who had had great experience in his occupation. In the midst of the intricate labyrinths of ice, turning sometimes to the right and sometimes to the left, now winding round a large hummock, now crossing over a smaller one . . . while I was watching the different turns, compass in hand, trying to discover the true route, he appeared always to have a perfect knowledge of it. . . . His estimation of the distances passed over, reduced to a straight line, generally agreed with my calculations, based on observed latitudes and the day's course. It was less difficult to preserve the true direction on a plain surface. To enable us to follow as straight a line as possible, we endeavored to keep our eyes fixed on some remarkable piece of ice at a distance; and if there were none such, we were guided by the wavelike ridges of snow (*sastrugi*) which are formed, both on land and on the level ice of the sea, by any wind of long continuance. These ridges always indicate the quarter from which the prevailing winds blow. The inhabitants of the *tundras* often travel to a settlement several hundred versts [one verst = 1.06 kilometers] off, with no other guide over these unvaried wastes than the *sastrugi*. . . . It often happens that the *true*, permanent *sastruga* has been covered by another produced by temporary winds; but the traveller is not deceived thereby; his practised eye detects the change, and, carefully removing the recently-drifted snow, he corrects his course by the lower *sastruga* and by the angle formed by the two. We availed ourselves of these ridges on the level ice of the sea, for the compass cannot well be used while driving; it is necessary to halt in order to consult it, and this loses time. Where there were no *sastrugi*, we had recourse to the sun or stars when the weather was clear, but we always consulted the compass at least once every hour.[24]

Tatarinow, the Cossack, seemed to be "guided by a kind of unerring instinct," which enabled him not only to find the correct direction while winding his way through the hummocky ice, but also to make the necessary detours compensate each other so that he was back on the course line afterwards, something that even von Wrangell, using his compass, had problems doing.

So how did the Cossack do it? I think I gave the answer in the previous chapter: He relied on the cognitive map that he made of the hummocky area. His "automatic mapping office" continuously gave him the information he needed: where he was and which way to go around the hummocks. He felt all the time in which direction he was going, if it was too much to the left or to the right, and also where he was relative to the course line he was trying to follow. Thus when he had to choose whether to go to the right or to the left of a hummock, he would "feel the pull" of the course line and return to it.

One can understand why von Wrangell, who tried in vain to keep track of the windings around the hummocks using his compass, thought that his driver must have had some kind of mysterious instinct.

When in an open area, the drivers used the obvious method, finding a recognizable ice formation that could serve as a landmark and keep them on course as long as they did not confuse it with a similar formation. In clear weather they also used the sun and the stars (at these high latitudes the azimuth of the sun and stars near the horizon changes approximately fifteen degrees per hour, but the spatial system can handle that).

They also relied heavily on what I call the "fossil wind compass" of the Arctic, the *sastrugi*, snow ridges formed by the prevailing wind and indicating its direction. It is as if the ground were covered with long compass needles pointing not to the north but in a direction well known to the experienced traveler.

An admirable description of Arctic way-finding, and one is especially grateful to von Wrangell for mentioning that his driver's uncanny ability struck him as "instinctive," which a modern scientist might have hesitated to mention for fear of being unscientific.

Darwin was the first one to comment on von Wrangell's report:

We must bear in mind that neither a compass, nor the north star, nor any other such sign, suffices to guide a man to a particular spot through an intricate country, or through hummocky ice, when many deviations from a straight course are inevitable, unless the deviations are allowed for, or a sort of "dead reckoning" is kept. All men are able to do this in a greater or lesser degree, and the natives of Siberia apparently to a wonderful extent, though probably in an unconscious manner. This is effected chiefly, no doubt, by eyesight, but partly, perhaps, by the sense of muscular movement, in the same manner as a man with his eyes blinded can proceed (and some men much better than others) for a short distance in a nearly straight line, or turn at right angles, or back again.[25]

Darwin omits the landmarks, the *sastrugi*, the sun and the stars, which for him must have been seen as trivial, while he, like von Wrangell, was fascinated by how the drivers managed to stay on the course line when they were winding their way through the hummocky areas.

Darwin hit many points right on the head: He is sure that the way-finding ability of the drivers is a trait we all have, differing in degree only, which I hope we can all agree on. He calls this ability "dead reckoning," which is a good way of describing it; knowing position by adding up the vectors of the displacements (the detours around the hummocks). He proposes that it is done "in an unconscious manner," which is also correct. He is quite right that eyesight plays a major role and adds a well-placed "perhaps" to his suggestion that a "sense of muscular movement is involved," the sense that enables some people to proceed in a straight line for a short distance when blindfolded.

Many other scientists commented on Darwin's account, and illuminated different aspects. H. C. Bastian shares Darwin's view that there is only a difference in degree, not in kind, between the sense of direction in humans and animals. He notes the big variation in our species from those city dwellers who often get lost, to the experts, the "semi-savages" and "savages," who come close to the prowess of the animals.[26]

C. Viguier in Algiers, unlike Darwin, does not believe that dead reckoning is involved. Instead he proposes that "since the magnetic state of the point to attain is known, the direction . . . will always be indicated regardless of the detours demanded by the difficulties of the terrain."[27] What he means is not the ability to feel the direction of the magnetic field, but a much more questionable ability to find the direction to your destination by comparing its known magnetic state with the magnetic state felt at your location. There are global positioning instruments capable of doing this, but no instruments based on the magnetic field. This, therefore, seems farfetched, indeed, but I dare not use the word "impossible." So many "impossible" things happen in animal way-finding, including the use of the earth's magnetic field, that caution is in order.

A fourth scientist, Pierre Jaccard, comments extensively on von Wrangell's story and on Darwin's discussion of it. He accuses Darwin of neglecting to mention the *sastrugi* that served as reliable direction cues, but stating instead that "the natives kept a true course . . . with no guide in the heavens or in the frozen sea . . . This last point was particularly important: all Darwin's theories about *dead reckoning* and as a result all the hypotheses on the sense of orientation were based on this alleged proof that there could be a purely inner direction frame in men as in beast."[28]

While Jaccard explains the simple part of the story, the guidance of the *sastrugi* over the open icefields, he does not explain the tricky part, the crossing of the hummocky ice with constant detours, that filled von Wrangell with admiration and Darwin with wonder. The reason for that is simple. He couldn't: it lay beyond his understanding.

Jaccard clearly suffers from intellectual hubris, looking

down on "savages" without education. Elsewhere he stresses that Tatarinow, the most skilled of the way-finders, was a Cossack and thus civilized. For him the possibility that the "primitives" could have a far better spatial ability than an educated Westerner is unthinkable.

Darwin did not have this self-imposed limitation. Neither did von Wrangell. Both of them had spent several years as explorers in close contact with primitive people and learned to appreciate and respect them as different but not inferior. They were both very well aware that they themselves possessed a tremendous amount of knowledge that the natives lacked, but they also felt that there were quite a few things that their Western culture hid from them—basic, important, vital things like way-finding when there was nothing to go by, that gave the natives the edge in their precarious balancing act on the brink of extinction. The natives showed them that there is a learning that cannot be acquired from textbooks and a knowing that cannot be taught by university professors.

CHAPTER 23

ABORIGINAL AND

UNDERWATER WAY-FINDING

I conclude this section of the book with three short stories that illustrate different aspects of our way-finding system.

David Lewis, renowned scholar of Polynesian and Micronesian wayfinding and author of *We, the Navigators,* also studied the natives in the Australian desert. He gives a good example of the dead reckoning system at work in way-finding, describing an experience from the *mulga,* an arid land in Australia where the dominant vegetation is a shrubby species of acacia. It is a difficult area for way-finding, with no landmarks and reduced visibility:

A kangaroo, wounded by a .22 bullet, was hunted on foot for half an hour. After it had finally been killed, Jeffrey and Yapa Yapa headed without hesitation directly back towards the Land Rover, that had been invisible since the first minutes of the chase.

Question: "How do you know we are heading straight towards the Land Rover?"

Answer: Jeffrey taps his forehead. "*Malu* (Kangaroo) swing
   round this way, then this," indicating with sweeps of his
   arm. "We take short cut."
Question: "Are you using the sun?"
Answer: "No."

The Land Rover duly appeared through the mulga in
about a quarter of an hour.[29]

From analyzing this story Lewis concludes:

It would appear then, that the essential psychophysical
mechanism was some kind of *dynamic image* or *mental
"map,"* which was *continually updated* in terms of time,
distance and bearing, and more radically *realigned at each
change of direction*, so that the hunters remained *at all
times* aware of the precise direction of their *base and*/or
objective.[30]

To which I can only say, amen. This is exactly how it is.
The cognitive map is continually updated; it has to be since
it is seen from one's own point of view, which changes continually as one moves. And since it is firmly held in place by
the direction frame, it will indeed be realigned each time
one changes direction. And in a featureless country the only
objects seen in the cognitive maps would be the base (the
Land Rover) and the objective (the animal killed), and
they would always be seen in the right direction as if the
mulga was transparent.
   Let me return to Colin Irwin, an outstanding witness

when it comes to spatial ability, since he himself possesses an excellent one (alas, not too common among those who study this problem). He describes his experiences in underwater orientation, a difficult environment for way-finding:

> During my late teens I was an active amateur diver around the coasts of Britain where the waters are never very clear. In the upper parts of the English Channel twenty-foot visibility was considered good; frequently it was down to a few feet. . . . However, some experienced people developed what we considered at the time to be a sixth sense as they were able to travel extensive circuitous routes underwater and always return to the anchor without backtracking, even in poor visibility. . . .
>
> A sense or rather knowledge of direction or orientation was given by ripples of the sand that ran parallel to the prevailing swell in shallow water or at right angles to the current direction in depths beyond the influence of the swell. . . . Additionally, the direction of drift of the diver across the seabed gave directions of tide and current. The surface boat would always lie on its anchor with the swell and/or current so that the diver on descending the angled chain would be given a directional "fix" at the onset of his immersion. Providing the diver could now discern "distance covered" and compute this against orientation he could travel over an extended area and return to the anchor. . . . The diver who is able to navigate "by the seat of his pants" using visual references from the sea floor will sometimes be able to navigate better than divers using conventional dead reckoning methods. . . .

I tried actively to improve my underwater navigation by making extended triangular exploration of several miles when I was in the Red Sea. At first my awareness of what I was doing was a hindrance as I tended to over- or under-compensate for navigational changes but in time my ability improved until I could travel extensively at night by watching the direction of the sand ripples in my flashlight beam.[31]

Let us look a little closer at his story. One thing is clear: unless there is some input that keeps the direction frame from deviating, the spatial system will not work. But in this case he gives a long list of directional cues available on the sea floor.

But then comes the tricky part, the dead reckoning. Irwin has to evaluate the direction and distance of all the displacements and add them up to get his actual position on the sea floor. And how does he do it? Well, he has a trick; he asks himself in what direction the anchor lies. One could express that differently: he focuses on the anchor in his cognitive map. Thus what he is actually doing is kicking in his "subliminal computer"—I call it the spatial system—and letting it take over. Using it, he learns to wander around for miles on the floor of the Red Sea at night without getting lost. Even with the sand ripples as a cue for direction this is an astounding feat.

It is an interesting observation that in the beginning his awareness of what he was doing marred his performance. It is natural when one is in a new situation to try too hard to get it right, and this interferes with the automatic pilot, just as I found in my desert experiment. One has to learn to relax and leave it all to the spatial system, which registers the directions walked in based on the visual input and the distances covered. It then does the calculations required and presents us with the readout showing the direction to the starting point. We must have complete faith in our "automatic pilot"; then all goes well. It is when we start to doubt that we get lost.

My friend P. L. Barton, in New Zealand, located the story of a Maori chief, or *tohunga*, that demonstrates this perfectly. The chief was asked to point out to a surveyor a boundary line from a point on the Waipuku River to the Kopua-tama pool.

Arrived at the starting point on the Waipuku, the surveyor waited on the tohunga's decision as to the direction of the line to be run. . . . After directing the surveyor to set up his instrument on the south bank of the Waipuku, the old tohunga recited an ancient karakia (incantation) calling on the atua, the spirits of the forest, to guide him aright in directing the line. This ceremony concluded, he stood by the surveyor and directed the clearing of the undergrowth in the general direction desired, and finally, after careful consideration, had a stake placed ahead of the instrument on the true line to Kopua-tama. The line was duly cut and run from this origin and after going straight across 8½ miles of dense unbroken forest growth the party finally came out on the western rim, a few yards off centre, of the pool Kopua-tama—a striking proof of the chief's sense of direction. The old Maori himself was fully assured that his success was principally due to the powers of his atua.[32]

I imagine most modern readers would regard the chief's preparation—contacting his *atua* through ritual incantation—as totally irrelevant. But is it really? When we keep in mind that the desired direction had to be retrieved from the unconscious spatial system as an intuitive feeling, it is clearly advantageous to eliminate everything that can disturb this process. I am convinced that if the chief had been worried about the outcome (and who would not be when asked to point to a fairly small body of water some 10 kilometers away) he would have been thrown off. The incantation ceremony lifted the responsibility from his shoulders to his *atua*'s. As a result, the chief believed he was no longer in charge of the proceedings, but just acted as instructed by his *atua*. His *faith* in the *atua* gave him the peace of mind he needed to perform optimally, and afterwards the chief humbly gave his *atua* the credit for his success.

At the level of consciousness where the spatial system works, illusions are not questioned but given the same full credence as reality! I speak from experience here: when my direction frame was reversed in Cologne, not even my knowledge that the sun had to rise in the east could make my spatial system forgo its illusion—I still felt that the sun was in the west. And as we shall see, this is not uncommon.

PART 4

WALKING IN CIRCLES
WHEN LOST

I remember a university professor telling me how he met his wife. The dean was giving a party for the faculty and the dean's wife was greeting the guests as they arrived with a few polite words for each one of them. He overheard the following exchange she had with a young female professor:

"I don't think I have met you before. Are you a newcomer?"

"No. I have been here for over a year now."

"Well, perhaps we don't move in the same circles?"

"I don't move in circles!"

That was when my friend decided that here was a woman he would like to share the rest of his life with.

But the fact is that when we get lost we do move in circles. It is attested in old folklore and in the scientific literature, and I was unlucky enough to experience it myself.

CHAPTER 24

THE *SKOGSNUVA*

FAIRY TALE

In his book *Wärend och Wirdarne*, "Värend (a county in the south of Sweden) and the Virds (its inhabitants)," Gunnar Hyltén-Cavallius writes about the legend of the *Skogsnuva*, or the female wood nymph. In the older Värend tradition she was seen as a forest troll, appearing as "a woman dressed in leather, dashing through the air with her hair all loose and long breasts thrown over her shoulders." However, in later tales she is "a fair maid, beautiful and seductive, when one sees her from the front; but when anybody sees her from behind, she is hollow at the back like a rotten tree stump."[33]

The Skogsnuva tries in every way to gain power over people who venture into the forest. He who walks in the wilderness must therefore be very careful. If he hears somebody calling him by name, he must not answer "yes" but "hello," for the Skogsnuva knows the name of everybody, and it might be her calling. If he answers "yes," she immediately gains power over him and at once lets out a roar of laughter resounding throughout the forest. When the Skogsnuva has gained power over people she confuses them in many ways. He who is led astray by the Skogsnuva can no longer find his way in the forest, but walks and walks, and always returns to the same point. Finally he becomes so dizzy and confused that he does not even recognize his own house. When that happens the only remedy is to turn the sweater inside out or say The Lord's Prayer backwards.[34]

I am not going to apologize for using an old fairy tale to make a scientific point. A story like this one is not created out of the blue, but probably starts out as several stories about similar occurrences, which are then fused together into one where the common element in the stories is enhanced. It is a distillate of the common experience of the society that developed it.

It is the custom when presenting the results of spatial ability experiments to first describe the subjects—usually university students lured into taking part by a small sum of money. Nothing is ever said about their spatial ability for the simple reason that nothing is known about it—except that they must have been able to find the laboratory where the experiments were performed.

The subjects in this fairy tale are anonymous. We know nothing about them except a very important fact: they probably had excellent spatial ability. They were forest people who roamed from childhood on in their very challenging environment, the coniferous forest. Their spatial ability was developed to its highest possible level, and never contaminated by the use of a topographic map or magnetic compass.

The hunter in the fairy tale makes several attempts to get out of the forest but always comes back to the same spot. He must therefore have gone in circles repeatedly while trying to go in a straight line. The cognitive map he makes of each attempt to get out of the forest is therefore a straight line. What happens is that his direction frame turns slowly around, misleading him.

When he has gone full circle and returns to the starting point there are two possibilities. If he does recognize the starting point upon returning, he feels he has suddenly been pulled back to it (Figure 27).

If his spatial system is very unyielding, however, he will feel so strongly that he is in another location that he is unable to recognize the starting point when he returns to it. Hence he will get a feeling of walking and walking through a never-ending forest (Figure 28).

In the first case he will conclude that some otherworldly evil power is pulling him back to the middle of the forest every time he gets close to the edge. In the other case he feels that the forest must be moving with him, pulled along by some malevolent force beyond his comprehension.

There are two important facts mentioned in the story that are definitely not inventions of fantasy. The first is that the wanderer who has gotten lost is unable to get out of the forest. He walks and walks but he always returns to the same spot.

Thus, in spite of the fact that he tries to get straight out of the forest, he is actually walking in circles. But all the time he feels he is keeping on a straight course, hence his direction frame is turning slowly around as he proceeds. When

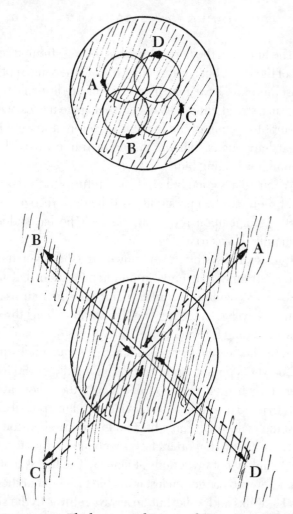

FIGURE 27, TOP. *The hunter makes several attempts to get out of the forest but always circles and returns to the starting point.*
FIGURE 27, BOTTOM. *The hunter's perspective: he feels he is going in a straight line. When he returns to the starting point and recognizes it, he feels he has somehow been pulled back to it.*

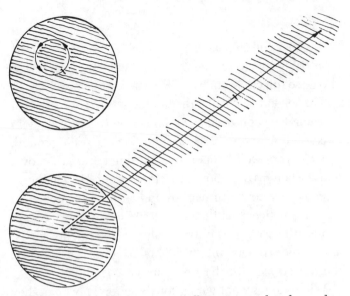

FIGURE 28, TOP. *The hunter actually goes round and round in roughly the same circle.*

FIGURE 28, BOTTOM. *The hunter feels he is going in a straight line. He has strange feelings of déjà vu, and therefore concludes the forest must somehow be following him.*

he has gone full circle and comes back to the same spot, he is probably able to recognize it, for his direction frame has also gone full circle and is back in the correct position.

The second fact is that the hunter gets so confused that he cannot recognize his own home. Since his walking in circles is due to the direction frame slowly turning round, it could very well happen that the hunter sees his own house when his direction frame is reversed. He will therefore see his house turned around 180° in his cognitive map. He notices that the house looks like his own, except turned the wrong way. Ironically, the better his spatial ability, the stronger will his feeling for directions be, and thus the stronger the barrier against recognition.

When things went wrong for these forest people, they blamed it not on themselves, but on some evil power, like the *Skogsnuva*, who had played a trick on them.

We find it rather ridiculous to believe that magic—turning one's sweater inside out or reciting a prayer backwards—could rectify the situation. But instead of being so cocksure, let us analyze this with an open mind.

We know that the hunter believes he is bewitched by the *Skogsnuva*. He therefore needs to change his identity. Since he cannot change his body, the only solution is to change his attire. He has no other clothes to change into, so turning the sweater inside out is the best he can do. And this will do it, provided he has *faith* that it will. For he is not trying to deceive the *Skogsnuva*; he is only trying to regain his self-assurance.

Reciting the Lord's Prayer backwards also seems to us at first sight like a silly and superstitious thing to do. But the hunter is in a state of panic, unable to function rationally. He needs to get out of that state to survive, a very difficult thing to do. His mind is running in a closed loop, repeating an ineffectual action again and again.

This backward recitation of the prayer could very well be just what the doctor ordered to break that vicious circle. It is a most difficult thing to do that requires total concentration. It takes about half a minute to recite the Lord's Prayer from beginning to end, and it can be done almost automatically, with no thinking at all. Which means that saying the prayer the usual way would be of no help to break the spell.

Reciting the prayer backwards is another matter. It took me about three minutes to do, and it certainly demanded my full attention. Any irrelevant thought sneaking into my consciousness was enough to throw me off, forcing me to start all over again. It is therefore an ideal means to force the hunter to concentrate on something long enough to remove panic from his mind. And the fact that it is the Lord's Prayer gives added power to it. He does not even need to be deeply religious to benefit, just deeply superstitious. And hundreds of years ago most people definitely were. Therefore, as long as he has firm faith that the spell is broken, it will be.

I am reminded here of the French scientist Pierre Bonnier, who lost contact with his environment when absorbed in reading.[35] His spatial system temporarily shut down during intense concentration. The same thing could happen to our fairy tale hunter when totally absorbed by the unusual mental task. With the panic eliminated, he would be able to evaluate his situation more calmly, and would have a better chance to find his way home again.

Unlikely as it may seem, many of the folk stories of supernatural occurrences in enchanted forests might have their background in situations such as this.

CHAPTER 25

GOING IN A CIRCLE

ON THE PRAIRIE

George Catlin, the well-known nineteenth-century painter
of American Indians, wrote:

> It is a curious fact and known to all the Indians, that the
> wild horse, the deer, the elk, and other animals, never run
> in a straight line: they always make a curve in their running,
> and generally (but not always) to the left. The Indian see-
> ing the direction in which the horse is "leaning," knows just
> about the point where the animal will stop, and steers in a
> straight line to it, . . . by a day's work of such curves, . . . the
> animal's strength is all gone; he is covered with foam, . . .
> his curves are shortened at last to a few rods [a rod = 5
> meters]. . . . I have said that the horse and other animals
> "generally turn to the *left*." How curious this fact, and
> from what cause! All animals "bend their course." Why
> bend their course? Because all animals have their homes—
> their wonted abodes, and they don't wish to leave them:
> but why bend to the *left*? . . .
>
> But man "bends his course"; man, lost in the wilderness

or on the prairies, travels in a curve, and always bends his curve to the *left*: why this?

While ascending the Upper Missouri . . . the vessel got aground, . . . I left the steamer with one man to accompany me, and . . . we started to perform the journey on foot. In our course we had a large prairie of some thirty miles to cross, and the second day being dark and cloudy we had no object by which to guide our course, having no compass with me at the time.

During the first day the sun shone, and we kept our course very well—but on the next morning, though we started right ("laid our course") we no doubt soon began to *bend* it, though we appeared to be progressing in a straight line. There was nothing to be seen about us but short grass, everywhere the same; and in the distance a straight line, the horizon around us. Late in the afternoon, and when we were very much fatigued, we came upon the very spot, to our great surprise, where we had bivouacked the night before, and which we had left on that morning. We had turned to the left, and no doubt travelled all day in a circle. The next day, having the sunshine, we laid (and kept) our course without any difficulty.[36]

We have here quite a bit of interesting information both about animals and humans. Catlin relies mostly on Indian lore when he states that animals do not run in a straight line—anecdotal evidence but given by people who knew what they were talking about. The Americans lived in symbiosis with the wild animals, and it was vital for their survival as a tribe to have a profound knowledge of the habits of the animals they were hunting.

He theorizes that animals "bend their course" because they are reluctant to leave their home. The effect is certainly that they return to the home base, which is generally an advantage for them, but one wonders if it is because they *want to*. It seems more like an instinctive response, advantageous for the survival of animals in the wild and therefore preserved in evolution.

Catlin then states that man also travels in a circle when lost in the wilderness, and always to the left. But this cannot be from any reluctance to leave the area. Man's home is *not* the wilderness, and—unless he is a fugitive—his wish is usually to *get out of* the wilderness and back to civilization. In Catlin's case it was certainly not his intention to return in the evening to the campsite he had left in the morning!

If we assume that Catlin and his friend walked five kilometers per hour and put in an eight-hour day, they walked a total of 40 kilometers (I walked that distance a couple of times in the Royal Hälsinge Regiment carrying more on my back than I like to remember) or a circle with a 6-kilometer radius. They must have walked in a very perfect circle in order to find their campsite again out on the monotonous prairie; it would be like looking for a needle in a haystack!

One cannot exclude the possibility that he unknowingly made several circles. He could very well have made two circles with 3-kilometer radii or even three circles of 2 kilometers each and been unaware of it. After all, he only became aware of having turned when he saw his campsite!

He states that the days before and after, when the sun was visible, they had no trouble keeping on course, and blames the cloudy weather for getting lost. Hence, at least one of them must have had a functioning "cognitive sun compass." One would imagine that on a treeless plain there ought to have been some wind on a cloudy day to help keep their bearings straight. Did they both lack a "cognitive wind compass"?

I found another similar story in Colonel Richard Dodge's book, *The Hunting Grounds of the Great West*, written in the nineteenth century. Here is an author with excellent spatial ability and therefore well worth reading even a century later for everybody interested in the sense of orientation. He and a friend went hunting in heavy fog, without their compasses. They decided to let the horses lead them home, but after several hours they came across their own tracks again.[37]

His interesting story confirms Catlin's report that animals can also end up going in a circle when visibility is reduced. One would have expected the horses to be able to find their way back home even in fog.

CHAPTER 26

GOING ASTRAY

IN THE CANADIAN

AND SWEDISH FORESTS

Since well-documented circle walks are difficult to find in the literature, I was happy to meet a person—I will call her Anne—who could describe the circumstances in detail and mark the route she had followed on a topographic map. It happened at the Head of St. Margaret's Bay west of Halifax. She was looking for Smelt Brook Lake in the early afternoon on a cloudy day in autumn. The woods in the area are relatively open; the visibility was about a hundred meters. She knew how to get to the marsh (see Figure 29), and afterwards she went north along the west edge of the marsh until she found a firebreak road that she followed westwards. When she reached a trail, she turned left and followed it in what she felt was the direction to the lake. Soon she was happy to see a ring of tall spruces ahead—thinking they showed the location of the lake—but when she came closer she realized she was back at the marsh instead. An interesting detail: she had walked the last part of this trail many times before, but still failed to recognize it.

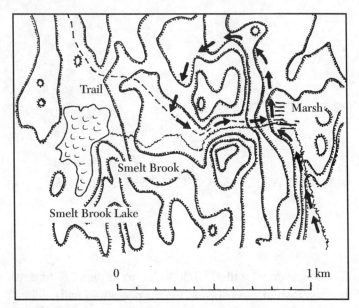

FIGURE 29. *The route followed by Anne when attempting to walk to Smelt Brook Lake.*

In Figure 30 I have reconstructed the route Anne felt she was following. It shows that when she reached the trail at A, her direction frame must have turned so much that she felt she had to follow the trail to the left in order to get to the lake. It also shows why she failed to recognize the trail in spite of being familiar with it. The reason was that the trail she thought she was following did not exist in the real world, only in her mind. There it was more than a kilometer to the west of the trail she knew, and it ran in the opposite direction. Thus her sense of location and her sense of direction, which normally help us recognize familiar areas, now worked against her and actually prevented recognition. When she passed familiar landmarks, they at most produced a strange feeling of déjà vu, nothing more. Compare her situation with that of the hunter in the *Skogsnuva* fairy tale, who got so confused that he failed to recognize his own house.

Only when she saw the very familiar marsh again did it dawn on her that she had walked in a circle. Instantly her cognitive map returned to reality and she jumped as if by magic from point B1 by the marsh in her illusionary map to the corresponding point B by the marsh when her cognitive map was still correct (Figure 30). At the same time her direction frame turned 180° back to the correct position. The spell was broken, and she was back in the real world!

Now let me describe my own circling experience. I'd had a lucky break, I would say, for it is one thing to read or hear about things happening to others, and another—much more instructive—to experience it oneself. I went hiking in the Swedish forest near my home town, Avesta, on a windy day with the sun only occasionally faintly visible. As usual,

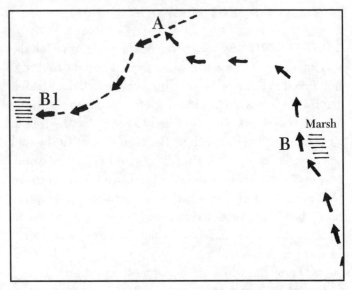

FIGURE 30. *The route Anne felt she was following. When hitting the trail at A she turned left towards where she felt Smelt Brook Lake was located. When recognizing the marsh at B1 she is immediately transported to its actual location at B.*

I had my compass in my backpack, in case I would need it. Not that I expected that to happen, since I had a good topographic map to go by.

But this time I was in for a surprise. I noticed it first when I got to the water tower. The road from the water tower runs towards the northeast (Figure 31), but I felt it was running southeast (Figure 32). Thus my direction frame was 90° degrees off.

At this point I decided to go on until I got to the road towards the north that would take me back to my starting point. I took off following trails all the time, trusting my sense of direction to take care of me. Finally I got to a road and turned right to go back towards Avesta. But there was something wrong with the road. It looked too new, it was too wide, and there was too much traffic. Reluctantly I got my compass out. It was turned around. At least I was smart enough to figure out that my compass was correct, and that it was my sense of direction that was turned around instead. What I felt was north was actually south. Thus instead of reaching the road I knew was north (Figure 32), I had continued veering to the left and actually ended up on the road south (Figure 31). Rather than take a short walk back along the road, I now had to retrace my steps along the trails I had come by to get back home.

As I write this nine years later, I can still visualize the road I ended up on. And, in my mind, it is still as in Figure 32—turned around and far north of its actual location!

It is amazing to think of what happened, how unsuspectingly I followed my sense of direction, despite noticing at the water tower that this very sense of direction was mal-

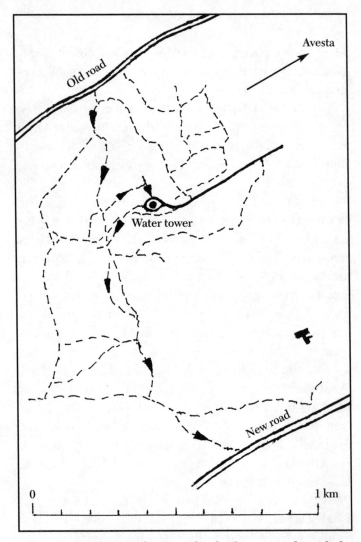

FIGURE 31. *A topographic map sketch of my route through the network of trails in the forest by Avesta.*

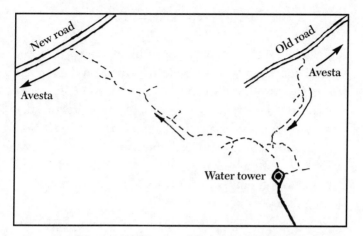

FIGURE 32. *The route I felt I was following. When arriving at the new road I mistook it for the old one. With my direction frame turned 180° I felt like turning right into that road to get back to Avesta, but this would have taken me away from the town.*

functioning. It shows what a powerful imperative it represents. The reason is of course that it works on the unconscious, intuitive level. Even when one is cerebrally aware that there is something wrong, as soon as one relaxes the conscious vigilance, the unconscious erroneous conviction kicks in automatically and leads one astray. The only way I could have gotten back to the old road, as planned, would have been to get my compass out and to follow it unquestioningly.

What happened also served as an eye opener. It showed forcefully what an asset this sense of direction must be normally, when it works correctly, when it automatically leads us to our destination. For it is only when something goes wrong, and it starts working against us, that we become aware of its awesome power.

I actually have a couple of excuses for the mishap. I had come by plane from San Diego—nine time zones away—a few days earlier, and was still suffering from jet lag. To top it off, I was very tired from the lack of sleep jet lag causes. The deterioration of the spatial system in old age (I was seventy at the time) could also have played a role.

It is important to note that *I did not go cross-country*. I followed trails all the way, except for some quite insignificant detours (only about 10 meters long) to avoid wet areas. *I deviated because I failed to evaluate correctly in which direction the trails I followed were running*. When I came to a trail junction I took the trail that my sense of direction told me was the right one. But, since my direction frame had drifted off, my feeling was wrong, and I took the wrong trail.

CHAPTER 27

HOW COME WE WALK

IN CIRCLES?

We now know that we walk in circles because our direction frame starts drifting steadily in one direction. The next question then becomes: Why does the direction frame, the mainstay of our spatial system, start drifting? The answer must be that the cues that normally hold it in place are lacking, or that they are present but for some reason the spatial system does not make use of them.

Then there is the intriguing fact that the direction frame does not move randomly back and forth; it drifts steadily in one direction. As Polonius says of Hamlet, "Though this be madness, yet there is method in't." So how do we explain this "method in the madness"? Well, those who ignore or deny that we have a "sense of direction," a spatial system with a direction frame, will have to look for the cause in the body. The body is not perfectly symmetrical. One leg is stronger than the other, one arm is longer than the other. Ergo, the reason we walk in circles is body asymmetry. When directional cues are lacking, the body takes over and leads us astray—or rather around.

But knowing that we *do* have a spatial system with a direction frame that keeps us oriented makes us wonder if it could be as simple as that. There is of course the possibility that body asymmetry *indirectly* causes the deviation by misleading the spatial system, by steadily pushing towards one side, causing a slight deviation that the spatial system fails to detect.

But I think that it is "all in the head." Let me explain: When we turn our head, this turn is registered by the horizontal semicircular canals (in the vestibular systems in our ears) and signaled to the spatial system so that it can keep our cognitive map oriented. For example, if you are sitting in a room facing a window and close your eyes, you will still be aware of the window's orientation because it is in your cognitive map. Now if you keep your eyes closed and turn your head, the cognitive map does *not* turn, and you are still aware of the correct direction to the window. Thanks to the input from the horizontal semicircular canals to the spatial system, the direction frame and cognitive map stay oriented regardless of how your head moves.

However, if there were an asymmetry in the system, if for example a turn to the right was registered by the spatial system as being somewhat larger than a turn to the left of exactly the same degree, this would lead to a slow drifting of the direction frame to the right. As usual, when one comes up with a good idea, one later finds that somebody else has already thought along the same lines, in this case two researchers in the field of human orientation, J. P. Howard and W. B. Templeton, back in the 1960s.

It is well known that after spinning about the body axis a person will veer in the same direction as the spin when attempting to walk straight. Slight persistent asymmetries in vestibular "tonus" could account for the consistency of the veering tendency. Such a cause might account for Schaeffer's finding that the mode of locomotion does not affect the direction of veering, and would fit in with his claim that the direction of veering is independent of structural asymmetries of the body.[38]

The reference to the article by Schaeffer (1928) sounded like pay dirt, so I got hold of it, and it turned out not only to be very interesting but also quite enjoyable reading:

In all motile organisms whatsoever, there is a deep-seated spiraling mechanism which leads an animal to go in a series of circles of spiral form when there are no guiding senses, as in the lower organisms, or when the guiding senses do not function through fear or for other reasons, in the higher.[39]

Schaeffer is convinced that walking in circles is not the exception in the animal kingdom, but the rule. Even higher animals fall back on it when the "guiding senses" that normally enable them to go on a straight course are knocked out.

His experimental results with blindfolded subjects rule out body asymmetry as a cause. For example, when told to go straight, the same subject sometimes veered to the left and sometimes to the right (Figure 33, left).

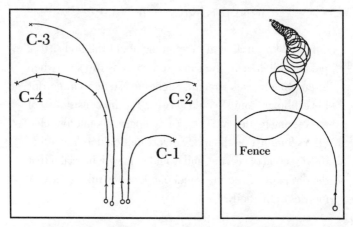

FIGURE 33. *Schaeffer's experiments on veering with blindfolded subjects rule out body asymmetry as an explanation.*

FIGURE 33, LEFT. *Subject C veering both to the right, C-1 (137 steps) and C-2 (140 steps), and to the left, C-3 (running slowly) and C-4 (walking 200 steps with 5 seconds rest after every 25 steps).*

FIGURE 33, RIGHT. *Subject C deviates slightly to the left. After 125 steps he is stopped by a fence, the blindfold is removed and he is told to return to the starting point. Again blindfolded he deviates to the left and after 80 steps he starts "spiraling" until the diameter is only 8 steps. (The figures are copied from Schaeffer's article.)*

In another experiment the subject (Figure 33, right) veered to the left, and hit a fence. The blindfold was taken off and he was sent back to the starting point. He set out again from the fence in the correct direction but veered to the left again and all of a sudden started spiraling in smaller and smaller circles until the radius got as small as four steps, or about three meters. For some reason the slow drift of the direction frame to the left sped up and finally rotated about four full turns per minute! Could it be that the stress of being deprived of visual cues for a long period resulted in a kind of sudden panic, causing the "brake" on the direction frame to collapse, and allowing it to turn faster and faster? Is this what happens when people get lost in a forest and panic and start walking in smaller and smaller circles?

Schaeffer was convinced by his experiments that the circling originates in the central nervous system and not in the legs. One would imagine that with such a body of experiments as support for his views, these would have become generally accepted. But instead they seem to have been suppressed by silence. The reason was probably Schaeffer's outspoken statements about other researchers in the field:

> The explanation [for why we walk in circles] is invariably based upon asymmetry of the body such as longer or stronger right or left legs, . . . The explanation for the phenomenon, though false, was so completely satisfying because of its simian simplicity that it must have seemed like wicked scepticism to investigate it by experimental

methods. This is an excellent example of the persistent devastating effect of substituting logic for experiment.[40]

"Simian simplicity," "wicked scepticism," and "devastating effect of substituting logic for experiment" must not have endeared him to his peers! Whatever the reason, Schaeffer's theory never caught on and the asymmetry explanation continued to hold sway.

I must confess I am suspicious of Schaeffer's use of blindfolds in his investigation. I see them as too unnatural and stress-inducing in these kinds of experiments. It ought to be better to use an area where the visibility is limited naturally, as in a forest that is fairly dense but still open enough for easy walking, and with no slopes that could be used for guidance. The experiments could be done under various conditions. One could test if most subjects functioned better on sunny days than on cloudy ones, or if a steady wind would improve the performance on cloudy days.

Since suitable forested areas are not always close at hand, one could make these experiments instead in large open fields at night with flashlights for guidance. That would limit the field of vision more evenly.

Come to think of it, one could create an artificial body asymmetry by putting a heavy hiking boot on one foot and a light shoe on the other. By running two series of trials, one with the left and one with the right foot heavier, one might be able to prove or disprove that the body asymmetry plays a role in the veering.

One might wonder how such a drawback as this tendency to circle could have survived in evolution. But if we look for example at Catlin's circling experience, one thing stands out: it would have been far worse if the deviation had been erratic instead of consistent. As it was, he ended up at his campsite and could make a fresh start the next morning. If he had moved erratically, at nightfall he would have been completely lost. It therefore seems plausible that consistent veering to one side could be an evolutionary safety net.

Take, for example, a deer being attacked by a mountain lion. The only chance of survival for the deer is to run away. If this was handled properly by the spatial system, the deer would keep running in a straight line. However, the deer would soon be outside its familiar area, and there it would lose the advantage of knowing the terrain. Even if it survived, it might have difficulty finding its way back to familiar territory.

It would therefore be advantageous for the deer if the part of the spatial system that, by detecting and correcting deviations, kept it on a straight course were disconnected. It could then try to outrun the mountain lion along a circular path and thus remain within its territory. This leads me to suggest the following hypothesis: We have a *tendency* to deviate to one side. Normally this tendency is held in check by the spatial system, which notices the deviation and corrects it. However, under abnormal conditions, e.g., fatigue, stress, or panic, the part of a spatial system that detects the deviation would be inoperative, our deviation would go unchecked, and we would walk in a circle.

PART 5

REVERSALS
OF ORIENTATION

"One doesn't notice the cow until the stall is empty," we say in Sweden, meaning we don't realize what we have until we lose it. This is especially true about the spatial system, which does its important job so unobtrusively that we are usually totally unaware of what is going on. It is only when something goes wrong that we suddenly wake up to its existence.

One such malfunction occurs when our sense of direction goes awry and we feel that north is where we know it is not and are unable to correct it. The most spectacular is the total reversal of the direction frame when we feel that north is in the south, but even 90° misorientations, when we feel that north is in the east or in the west, can be most annoying.

Since it is when the black box hidden in the unconscious mind cracks open that we have a chance to peek inside, these events are manna for the student of the spatial system, because when we are able to find the cause for the misorientation we get an indication of what goes on when the system works properly, which it luckily does practically all the time.

CHAPTER 28

FORDE'S LETTER

TO THE EDITOR OF *NATURE*

It was Charles Darwin who, in 1873, got the ball rolling on the study of directional misorientation in his article "Origin of Certain Instincts," published in *Nature*:

> The manner in which the sense of direction is sometimes suddenly disarranged in very old and feeble persons, and the feeling of strong distress which, as I know, has been experienced by persons when they have suddenly found out that they have been proceeding in a wholly unexpected and wrong direction, leads to the suspicion that some part of the brain is specialised for the function of direction.[41]

This triggered a letter to the editor from H. Forde about the subject; a letter that created somewhat of a stir in the scientific community at the time.

> In Mr. Darwin's article in NATURE for last week there is a passage about "the sense of direction being sometimes suddenly disarranged," that brought to my mind assertions I had frequently heard made when travelling some

years back in the wild parts of the State of Western Virginia. It is said that even the most experienced hunters of the forest-covered mountains in that unsettled region are liable to a kind of seizure; that they may "lose their head" all at once, and become convinced that they are going in quite the contrary direction to what they had intended, and that no reasoning nor pointing out of landmarks by their companions, nor observations of the position of the sun, can overcome this feeling; it is accompanied by great nervousness and a general sense of dismay and "upset"; the nervousness comes after the seizure, and is not the cause of it. . . . The feeling is described as sometimes ceasing suddenly, and sometimes wearing away gradually. Would it not be strange if it should appear that there is a sense of direction other than an acquired sense of direction the result of unconscious observation, and that some animals possessed the first in a pre-eminent degree?[42]

Let us see now, one and a quarter centuries later, if we can figure out what went on in the head of these unfortunate hunters. The story is sketchy, but with careful analysis we can still find enough hints to create a fairly good picture.

Forde states they experienced "a kind of seizure," and that "they . . . all at once . . . become convinced that they are going in quite the contrary direction to what they had intended." This makes it clear that what happens is a sudden reversal of their direction frame. Another important fact is that "no reasoning . . . nor observations of the position of the sun, can overcome this feeling." This shows that the direction frame turns around from a correct to an incorrect position. This is exactly what happened to me in Paris, where I had a "seizure," a most unpleasant sudden reversal of my direction frame, every time I returned to my hotel in the Château d'Eau area.

It is thus clear that the hunters entered an area where they already possessed a reversed cognitive map, and that it was seeing the landmarks in this reversed map that forced the direction frame to turn around suddenly (Figure 34). The hunters therefore must have been to the place before and at that time encoded a reversed cognitive map. The reason the cognitive map of this place became reversed during the first visit was probably that their direction frame came adrift while they were walking towards the place and had turned some 180° by the time they reached it. Since they had a good sense of direction—attested by their strong discomfort after the "seizure"—it is a good guess that the sun was hidden by clouds during this first visit. It is also likely that fatigue, both physical and mental (from encoding a cognitive map during a long walk in a new area), played a part.

Due to their strong sense of direction, their reversed cognitive map could not be corrected and the reasoning of the other hunters fell on deaf ears. The sun therefore stayed in the wrong direction as far as they were concerned, just as it did for me in Cologne. Only when they got away from the area where they had encoded a reversed cognitive map would they see the sun in its proper place again. This is usually a gradual process, unless they enter an area with a previously encoded correct cognitive map, which will instantly force the direction frame to turn back to the cor-

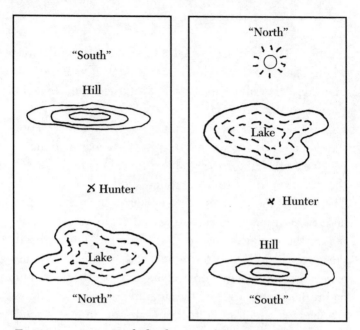

FIGURE 34, LEFT. *With the direction frame turned around on a cloudy day, the hunter encodes a cognitive map of the area, in which what he feels is to the north is actually to the south.*

FIGURE 34, RIGHT. *Entering the "turned-around" area at noon on a sunny day, the hunter experiences a sudden fit of "spatial vertigo," during which he feels the whole universe spin around. Afterwards he feels that the sun is in the north.*

rect position. I am sure when they got reoriented again, they would have heaved a sigh of relief, as I did in Cologne.

A later scientist, Viguier, pointed out in 1882 that:

> [T]his kind of direction vertigo is absolutely unexplainable by the theory of unconscious observation; for the result of observations made unconsciously with the help of various senses cannot be subject to this sudden and temporary beclouding. In the case, on the contrary, when we assume a special sense organ, it can be subject to momentary illusions, like the eye and the ear.[43]

He then concludes "that there exists, not only in animals but also in man, a true sense of direction; a sense whose keenness varies greatly according to the subjects, and . . . no doubt is located in an organ just as distinct as that of vision and hearing."[44] He finds the most likely candidate to be a magnetic sense, and I agree.

He thus was not able to explain what happened to Forde's hunters, nor was anybody else. Just reading about something like this happening to somebody else is not enough of a key to understand what was going on. I am sure Darwin himself read Forde's letter closely but was unable to explain the strange happenings. I understood it because I was lucky enough to experience exactly the same thing— albeit in Cologne and Paris instead of in the forests of West Virginia—and could therefore analyze it based on my own observations.

Some more examples of reversal were noted by Alfred Binet, the French psychologist, in his article "Reverse Illusions of Orientation" of 1894. He wrote about the observations he had collected: "they come from persons whom I see daily, whom I have long known, and in whom I place all confidence. All these persons have habits of close observation."[45]

In one of those observations Jacques Passy, a French chemist, reported the following childhood experience:

> I remember that while a child, returning home one day from the Bois de Boulogne, I asserted to those that went with me that we were following a road exactly opposite to the right direction, and I found the direction that they forced me to take absurd to such a degree that I wanted to cry, and my error disappeared only when near the house. At the end of a certain period of conflict, three or four minutes perhaps, the illusion disappeared. And when it disappeared for one street and one object it disappeared for all. I could then immediately take the road [=find my way] and direct [=orient] myself correctly.[46]

When in the Bois de Boulogne with its winding roads he must have become misoriented so that when they returned home he was utterly convinced he was going away from home instead (that is, his direction frame must then have been about 180° off). Only when he recognized the familiar area near his house did he get his bearings straight again. He describes the process as a conflict, which is a good way of seeing it, since it is a conflict between two cognitive maps colliding, the new reversed map he carried with him from the Bois de Boulogne, and the old map he had of the area around his house. After a couple of minutes the old map, on the strength of all the well-known landmarks in it, won the tug-of-war.

He then states: when the illusion "disappeared for one street . . . it disappeared for all," which he apparently finds rather puzzling. But it is easy to explain. When the illusion disappears, i.e., when the direction frame turns back into the correct position, it is not only the street we see that turns around, it is the whole universe.

We might find it strange that he is able to describe this childhood memory so well. After all, it must have happened at least twenty years before he wrote the report about it to Binet. The reason is that these experiences have a profound impact; they touch at the very foundation of our being, at the ties that join us to our environment. When things go wrong at that level we are really shaken up; it is not something we easily forget.

He notes that "the imaginary direction was exactly opposite to the real direction" and cannot "recall having [had] an impression of direction at right angles to the real direction."[47] The reason for this is that when one goes out of a house into a familiar street, there are only two directions to choose from, either the correct one or the one directly opposite. If the direction frame has slipped less than 90°, one will take the correct direction and, as one follows the street, the direction frame will quickly orient itself. However, if the direction frame has slipped more than 90°, one will take the wrong direction and as one proceeds, the direction frame will settle down in a position 180° off. This explains why Passy never experienced a 90° error.

Binet did, however, himself experience a 90° error. He saw a street in the distance and thought it was a familiar one running north–south, when, in fact, it was another street running east–west.[48] But in his case the reason was that he mistook a street for another one running at right angles, not that he felt that a familiar street was turned around.

CHAPTER 29

STRANGE MORNING

AWAKENINGS

In the same article, Alfred Binet describes another French scientist's study of awakening in the morning with his sense of direction reversed.

> M. Flournoy, in his recent work *Les Synopsies* (p. 188), speaks incidentally of these phenomena, which he has observed in himself in the moment of waking. "Who has not happened," says he, "to waken in the darkness of night with the curious idea that the room is turned in some way. We recall ourselves, and know certainly that we are lying in bed with the right side towards the wall, and in spite of this we feel that we have the wall to the left and the room to the right and are amazed at the tenacity of this illusion, which remains for many seconds in spite of reason; until, extending the arms, the contact with the wall causes it to suddenly vanish and brings the mental images back in their proper position. It is to be presumed that this reversal at awakening from sleep is due to the prolongation of a dream in which one has thought himself in another chamber.[49]

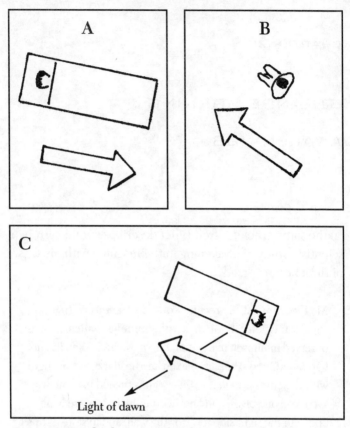

FIGURE 35 A: *The direction I faced on my camp bed when sleeping.*

FIGURE 35 B. *The direction I felt I was facing when dreaming.*

FIGURE 35 C. *The direction I felt I was facing when waking up and seeing the light of dawn.*

This shows that Flournoy thought most people have awakened during the night with their bearings reversed. He is probably generalizing from his own experiences; I know it happens very rarely in my case. But people have told me that, to them, it is a common experience. This is a reminder that one must be careful not to think, "If it happens to me, it must happen to everybody." There are great individual differences in human spatial ability.

The explanation offered that it could be due to a prolongation of a dream, where the person imagines himself in a room with a different orientation, is sensible enough. The only way to know if this is the case would be to ask people who experienced it as well to remember the dream they had just before waking up. This is not an easy thing to do.

I once experienced such a misorientation after a dream. I was sleeping at a campsite on Sierra San Pedro Mártir in Baja California the night of July 7. With my head propped up by the pillow I was looking east-southeast (Figure 35A). Just before waking I dreamed I was riding a bike at night without a lamp (I remember worrying about being caught by the police) and afterwards I rode in a train. The whole time the movement, as I perceived it in the dream, was west-northwest. I was thus turned around about 180° in the dream (Figure 35B).

When I started to wake up, I was on my left side and my first impression was that I saw the light of dawn over the southwestern horizon. I was thus turned around approximately 180° just as in the dream (Figure 35C). When I woke more fully and recognized my surroundings, this impression disappeared easily and without any sudden turnaround as I experienced in Cologne and Paris.

I certainly realize that this does not prove anything. All one can say is that it is a vague indication that directions in dreams can be carried over into the awake state afterwards, provided—as one may assume—that the dream takes place just before awakening.

Victor Henri, a science student in the late nineteenth century, describes a somewhat different illusion of orientation upon awakening experience:

Second Observation. (Made on the morning of January 1, 1894)—In my chamber the bed is in front of the window. I have to lie on my *left* side to see the window. The curtains are thick and let in very little light. In the evening I lay on my left side, and I do not recall having turned during the night. I slept well. In the morning I awoke, opened my eyes, tried to see if it was day and looked fixedly before me so as to see the window. But I could see nothing at all, not a trace of light. Then *suddenly*, I had the vague feeling that something was wrong, only I knew not what. I said to myself, "I am lying on my right side: the door is there (I indicated the opposite direction), so that I ought to see the window before me; but I do not see it." Then I made a movement with my hand and touched the wall. I felt it with astonishment [expecting it to be on the other side of the bed], recognizing that I was very much deceived. Turning, I saw the window on the other side, and I felt at the same moment that I had deceived myself as to the direction of the door.

During the night I had turned over without knowing it. The result was the same, therefore, as if the bed had been turned 180°.[50]

Henri's explanation that the reversal took place because he had been lying on the left side in the evening, and then turned over during the night without being aware of it, so that he was lying on the right side in the morning, is obviously wrong. In order for the result to be the same as if the bed had been turned 180°, he would have had to get up and lie down again with his head where his feet had been. What had happened was instead that his direction frame, and with it his cognitive map of the room, had been turned around 180°. This made him feel that the door was at the wrong end of the room, the window on the wrong side of the bed and the wall in the wrong direction (Figure 36).

As in many of those stories, the subject, who is trying to describe what went on in his mind when it happened, is having problems expressing himself clearly. Introspection is never easy and in this case, when there are two different cognitive maps of the same environment to sort out, it is natural to be confused. And the experience is so strange and has such a profound impact, especially the first time, that one is dumbfounded and unable to give a coherent report afterwards.

As for the cause of Henri's nocturnal turnaround, I wonder if the date could not have had something to do with it. The morning after a traditional New Year's celebration is

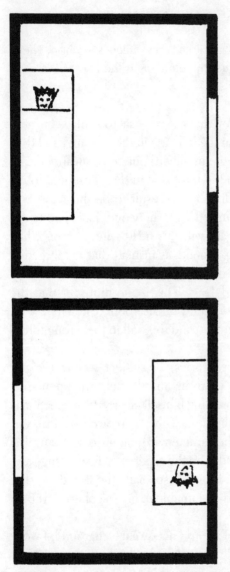

FIGURE 36, TOP. *The actual orientation of the room.*
FIGURE 36, BOTTOM. *How Henri felt the room was oriented when he woke up.*

often characterized by a significant hangover. (I recall from my bachelor days waking up after a similar occasion and finding, to my dismay, that my bed was moving like a ship in a storm.)

CHAPTER 30

WHEN TRAINS TAKE OFF

IN THE WRONG DIRECTION

In this chapter I examine three reversals of the sense of direction during train rides.

The first example is an observation made by Mr. Courtier, a demonstrator in the Psychological Laboratory in Paris, as related by Binet.

> One night I took the train from Rouen to Paris. At the end of a half-hour I fell asleep. At Vernon I suddenly awoke and, going out, walked a few moments on the platform of the station. After entering the car again I had the sensation of returning to Rouen. It seemed as if we were going away from Paris, although I took the same seat. There were at the station, the instant when I descended, two trains from opposite directions; and although I had not crossed the track, and although I did not think I had crossed it, I asked myself for one or two minutes whether I had not been deceived in the train [if I had not taken the wrong train]. I was somewhat anxious, and found it necessary to ask my neighbors as to the fact. When I had been assured that we were going toward Paris, the illusion still persisted for

some minutes. It was half past nine in the evening and very dark. I have taken this trip a hundred times in ten years, and have suffered the illusion but this once.[51]

The next observation is by Mr. P. Thélohan, who was Head of the Histological Laboratory at the Collège de France:

I have often felt a complete change of direction on the railroad or in a carriage. It has been generally on waking from a more or less profound sleep, or after some degree of drowsiness, that I have had this illusion. For example, being asleep in the car, I was at a given moment awakened by the arrival of the train at a station, and by the noise in the depot. Then sleep began to seize me. The departure of the train caused an awakening more or less complete, and it then seemed to me that the direction of the journey was absolutely changed. I remember very clearly having had under these conditions a moment of very disagreeable anxiety, of real pain. It was only after some moments that, after thoroughly waking, I recognized my error, by comparing the relative situations [locations] of the different objects.[52]

My own experience happened on April 1, 1993 (April Fool's Day, no less), in a commuter train from Sartrouville, a town on the Seine, to the Gare St. Lazare in Paris. I wrote what had happened in my notebook immediately after.

When I took the train to Gare St. Lazare, I had to stand at first. Then a lot of people left at one station and a lot came on and I managed to find a seat in the bustle. When the train started up again, I had the strong feeling that it was running back to Sartrouville. This impression was still with me after about five minutes, so I had to argue with myself to keep oriented. It was as if, when I stopped the conscious effort to keep oriented, it automatically slipped back in reverse. It was as usual difficult to get a firm grip on this. After perhaps ten minutes, I felt it was going the right way. REASON FOR THE REVERSAL: 1) It was cloudy, with no indication where the sun was; 2) I had only slept 3–4 hours, so I was exhausted; 3) I still suffered from jet lag after arriving from San Diego two days earlier, and at that moment the time was 11 P.M. in San Diego.

During the following days, I continued to think about this turnaround, which corresponded to those described by Binet. When I had read them earlier I had an uneasy feeling of fumbling in the dark. They did not resonate in my mind; the introspective analysis that can be done with one's own experiences was impossible.

It was therefore with mixed feelings that I contemplated what had happened when I felt the train going the wrong way. On the one hand, my confidence in my spatial ability was shaken, and it hurt my pride that my spatial system could have made such a mistake. On the other hand, I was happy to get direct personal experience of the spatial system malfunctioning in this way.

I think the main reason that it happened to me was extreme tiredness. I was seventy years old at the time, and experiencing jet lag after arriving from San Diego—nine hours' time difference—two days before, which I suspect can be very detrimental to the spatial system. I was also suffering severely from lack of sleep. The cloudy, misty weather with rain hanging in the air certainly did not help. It was impossible to see the sun that day. There was thus no input from that important cue for directions.

I think it is interesting that I was able to orient myself by making a conscious effort. It was as if my conscious mind could overrule my spatial system. But when I relaxed my conscious vigilance, the spatial system reasserted itself and I slid back into the feeling that the train was running the wrong way.

It seems that the gradual fading away of the reversal is not a continuous process. Instead, there is a period when the direction frame can move back and forth between reversed and oriented; one moment the conscious mind rules and it is oriented, the next the unconscious mind bounces back and it is reversed. Finally the spatial system, with no reversed landmarks to hold on to, capitulates, and the reversal disappears for good.

The turnaround must have happened while my attention was distracted in the bustle at the station before I sat down. Perhaps in the melee my direction frame "lost contact" with the carriage, so that afterwards, when the commotion had stopped, it could lock into the reversed position. Or perhaps my tired spatial system gave up keeping track of my turns as I was pushed around and just shut down temporarily, waiting for conditions to stabilize. Whatever the reason, there must definitely have been some kind of slippage in the normally firm connection between the spatial system and the environment.

My age and my mental tiredness must have made it easier for the turnaround to happen, and must also have made it easier for my mind to overrule the spatial system that held on to the illusion. I cannot recall something similar ever happening to me before, which makes it likely that my age was an important factor.

In the previous chapter I dealt with reversals upon awakening, some seemingly caused by dreams. It could be that such "misoriented dreams" played a role in the Courtier and Thélohan reversals. However, I did *not* sleep in the train, so in my case the dream hypothesis is out, but I *was* extremely tired. Could be that it is the *tiredness* that is the determining factor; that tiredness is in itself sufficient to cause the spatial system to malfunction, to slip; that therefore we need not look at more unusual situations—to reversals in dreams just before waking up—for explanations?

So this chapter ends with a question, and since experimental data on dreams are hard to come by, it will be difficult to find a definite answer.

CHAPTER 31

INDOOR MISORIENTATION

We have seen that an absence of directional cues, due to darkness, sleep, exhaustion, or other conditions, can cause misorientation. A lack of cues indoors, where there is no sun or landmarks, can have the same effect.

From 1880 to 1882 Binet often visited the Louvre museum but never paid attention to directions as he walked around, concentrating instead on the works of art. Since most of the rooms were lit from above, there were few cues for determining direction. It would therefore have been difficult for him to stay oriented. Once he happened into a hall with windows that faced the Seine embankment. There a strange thing happened. He writes:

> I saw the Seine rolling before me from left to right; but it seemed quite wrong, for in the position in which I found myself the Seine ought, as I thought, to roll in the opposite direction: the landscape seemed to be turned around. This was a common experience. [It happened every time.] Often, after several intervening days, I have felt this particular illusion reproduce itself, accompanied by a painful sensation. Then I ceased to frequent the Louvre, and I did not have another opportunity to observe this. A year ago I

tried again. Going to the museum I made no effort to keep my direction, and after an hour of walking approached a window opening into the Quai de Seine [Seine embankment], but the illusion was not produced.[53]

Binet walks around in this enormous museum concentrating on the paintings and sculptures the way a good visitor should. His mind is totally absorbed in the art while his spatial system makes a cognitive map of the place, the way a good spatial system should.

But humans are not designed for indoor orienteering. There is no sun to help (he points out that the skylights give no indication where the sun is), no wind, no general slope of the land, and without those cues one's direction frame might start drifting off, as it did in this case, by 180°, totally reversed. I do not think the angle had to be exactly 180°; to be precise, the direction frame must have drifted more than 90° and less than 270° when he looked out of the window and saw the river. Let me explain with an example.

Assume that the direction frame is 120° off when he sees the river. What will happen? Well, first we can come up with a good guess as to what will not happen. Binet has a general cognitive map of Paris, in which the direction of the Seine is well defined as it flows past the Louvre, from east-southeast to west-northwest. He could not possibly see it as flowing due south, as his 120° misoriented direction frame would want him to do. It is not strong enough for that. Instead it turns to accommodate the new information that the sight of the river gives.

For this direction frame turn there are two possibilities.

Either it turns 120° so that he sees the river flowing correctly west-northwest, or it turns 60°, so that he sees the river flowing in the wrong direction, east-southeast. If he does not immediately notice which way the river is flowing, the direction frame will most likely turn through the smaller angle and he will see the river flowing "upstream."

Thus the direction frame does not have to be exactly 180° off for the reversal to occur. The closer the deviation of the direction frame is to 180° when a linear landmark with known orientation is observed, the higher the probability that a reversal will occur.

You, the reader, might protest that it must have taken only a few seconds to see which way the river was flowing, and this ought to have corrected the mistake. But I strongly suspect that once the spatial system has made a faulty decision as to directions and made a false cognitive map inside that erroneous direction frame (which probably takes only a split second), then it is too late for the decision to be overturned. For now we have only flimsy anecdotal indications to go by, but if anybody should make proper experiments to prove or disprove my hypothesis, I am confident it will stand.

Actually the shape of the landmark does not need to be elongated, it is enough that it looks roughly the same when seen from the opposite side. In Paris, there is the Arc de Triomphe, in the center of what was formerly called Place de l'Étoile, that is symmetrical around two axes. Thus from wherever you look at it, there is an opposite direction from which it looks basically the same. I learned this fact the hard way when a friend and I were driving through Paris. We came up the Champs-Élysées and were going to turn left at the Place de l'Étoile, which meant going almost a full circle around the arch. With twelve streets feeding into the Étoile and traffic a nightmare, I lost count of the streets before we reached the one we wanted to turn into. I tried desperately to reorient myself by looking at the arch but soon realized that it was too ambiguous a landmark for directions. I do not remember how many times we went around this nerve-racking "merry-go-round" before we finally got out of it—not where we had intended, but alive, which at that moment was our main concern.

Then there is the Eiffel Tower, which looks the same from four directions. Here it is possible on a foggy day—or in a foggy mind—to get turned around not only 180° but also 90°.

I had better point out here that Binet's cognitive map of the Louvre served him well. He never got lost, or had any problem getting out again. In fact, if he had not happened to look out of the window and see a familiar landmark, he would never have noticed that he was misoriented. I already stated that we are not designed to keep our bearings straight inside buildings, but luckily there is seldom any need to. If we suddenly decide we want to go to another place, we are not going to make a hole in the wall and take off in the proper direction; we exit first and then, when we are out in the open again, we orient ourselves and find out which way to go. So all we need indoors is a cognitive map good enough to take us back out again, and even a rather distorted one can handle that.

Binet writes that he had a painful sensation when he

looked out of the window. This is understandable when we think of what he saw: the Seine to the *north* of the Louvre and, even worse, flowing *upstream*—a very disturbing sight indeed for any Parisian. But the pain could also have been caused by a kind of vertigo due to his direction frame turning to align with the river, which could very well cause some discomfort.

He tells us that he did not get turned around when he came back to the Louvre eleven years later. The reason for this is that, in the meantime, he had forgotten enough landmarks in his cognitive map of the interior of the museum so that he now made a completely new one that was correctly oriented with respect to the outside world—or, to be accurate, one that was so close to correct, i.e., less than 90° off, that it did not get turned around when he looked out the window. The changes in the exhibitions during those eleven years in which he did not visit the museum made it easier for him to forget the old erroneous map.

My own experience supports this. I actually went back to my old reversal area by the Château d'Eau métro station in Paris about forty years after the reversal had happened. I expected my reversed cognitive map of the area to remain. But this did not happen. Instead I encoded a new, oriented cognitive map of the street. Apparently my old map was too faded to influence the spatial system.

This compares with my friend Virginia's experience. She got misoriented 90° clockwise when arriving at night in Berkeley to study at the university and this misorientation remained for years. After finishing her studies she left the place and returned to visit again only some thirty years later. She then found that her old misorientation was alive and well. The reason for this is probably that the engraving of her misoriented map must have been much deeper than my Parisian map since it went on for years, while my visit in Paris lasted only a week.

Thus, the fact that we have an old reversed cognitive map of an area does not mean that we are bound to use it when we return there. If the time interval is very long, the old map will fade so much that we make a new, correct one next time we go there.

CHAPTER 32

ANALYZING MISORIENTATIONS

Now that we have examined the different kinds of misorientations, we can compare Binet's century-old analysis
with our current explanations. I have the highest regard
and admiration for Binet, but he wrote a whole century ago
and we have learned a few things in the meantime.

> We can distinguish three cases: (1) normal orientation—in
> which the points of reference [landmarks] recognized con-
> firm the former sense of direction; (2) disorientation—
> one has no sense of direction at all, and if he meets a
> familiar point of reference [landmark] he accepts it and ori-
> ents himself properly; (3) inexact orientation—one meets a
> point of reference, finds it in contradiction with his earlier
> system; the false system persists, even though he knows it
> to be false, just as an illusion persists. This last is the case
> now under discussion.[54]

Binet thus recognizes three "operating states" of the
spatial system, one where the sense of direction is right (1),
one where it is wrong (3), and one where it is in limbo (2).
State 2 is transitory; when we find a landmark and recognize it correctly, we get the proper sense of direction and

end up in state 1. However, if we misinterpret the direction of the landmark we get misoriented—turned around—and we end up in state 3. In short: in state 1 we are oriented, in state 2 we are disoriented, and in state 3 we are misoriented.

The amazing thing about state 3 is, as Binet points out, that if we run into a familiar landmark while we are turned around (e.g., we see the sun coming up in the morning in the "west"), this knowledge does not change our false orientation. State 3 is interesting because it gives us glimpses of what goes on inside the "black box" deep in our unconscious mind that provides us with our spatial ability.

As we have seen, the misorientation illusion can vary in strength. Sometimes it is weak; familiar landmarks are recognized, but we have a feeling that they are turned the wrong way. And sometimes it is so strong that landmarks give a feeling of déjà vu, but we cannot recognize them because they are turned the wrong way.

The stronger our spatial system, the more vivid the illusion when it is turned around. It follows, ironically, that the better our spatial ability, the more of a nuisance it will be when it turns against us. We are likely to fail to recognize familiar landmarks because they seem turned in an unfamiliar direction.

As previously discussed, most misorientations are 180° because when we run into a familiar linear landmark while deluded, there are only two possibilities: Either we get it right, or we end up turned around 180°. For example, you might make a mistake on a cloudy day and jump to the conclusion that the south end of a known street is in the north.

You cannot possibly conclude that the south end is in the east, since you know the street runs north–south.

While Binet believes that the illusion can be created by either conscious or unconscious judgment, as I see it, the decision is definitely made on the unconscious level; that is why we cannot later overturn it. However, before the decision is taken, all available clues are probably weighed against each other and at that time a conscious false conviction can be decisive. For example, when I saw Cologne across the Rhine at night and was convinced I was on the west bank of the river when in fact I was on the east bank, this conscious judgment most likely caused the illusion that north was south. But when I came out of the métro station in Boulevard de Strasbourg in Paris, it is more likely that I was already turned around because of unconscious judgments during the subway train ride and the subsequent walk through the corridors towards the exit. When I got out in the street I then unconsciously aligned it with my already turned-around direction frame, which gave me the illusion that north was in the south. This kind of analysis is a tricky business, because it is often impossible to say what part of the decision-making was unconscious.

But Binet also notes that in several cases "the observer took no pains to orient himself. The orientation was made automatically, while he was occupied with something entirely different." He correctly interprets his experience in the Louvre to be an example of that. "During my promenades in the halls I had made an unconscious orientation of the objects." He had made a cognitive map of the museum where he unconsciously localized the north in a wrong direction without thinking about it. This did not bother him until he looked out of the window and realized that Paris had turned around!

What puzzles him is when the illusion appears suddenly without any apparent cause, as in one of his own stories where he was sitting in an omnibus and suddenly felt it was going in the wrong direction despite his knowledge that it could not have turned around. He could give no explanation.

In those cases I think there is a kind of slippage in the system during a temporary shutdown, when we lose contact with the environment so that the system can start up again in a different orientation. This shutdown is probably caused by tiredness and/or intense concentration of the mind "so that we forget everything around us." I wonder if that is not what Darwin had in mind when he wrote how "the sense of direction is sometimes suddenly disarranged in very old and feeble persons." It might be that it is the *tiredness* that comes with old age and deteriorating health that causes the slippage and not the age and health, per se.

Binet ends his paper with the following:

> We still need to know whether the illusion is produced or not by a particular derangement of one sense-organ—possibly the semicircular canals of the inner ear. In experimental studies that have been made on the sensation of vertigo, no one has, to my knowledge, produced such illusions of the orientation of objects. The question is interesting and certainly deserves more study.[55]

The vertigo that we feel when spinning around, or just after the spinning stops, is caused by an "overloading" of the horizontal semicircular canals in the ears. They signal to the brain the wrong speed of turning, or that we go on turning when we have already stopped.

The vertigo that we feel when the direction frame suddenly spins around is something entirely different. Both the eyes and the semicircular canals correctly signal to the brain that we and our environment are stationary. The spinning around takes place entirely in our spatial system. The eyes, when observing landmarks that do not fit inside the direction frame, only give the input that triggers the action, forcing the direction frame to turn around to fit the landmarks in the cognitive map.

This "revolution" deep in our unconscious mind that we are unable to understand and control leaves us bewildered, shaken, and anxious. We realize to our dismay that our deepest need, a stable environment, can no longer be taken for granted.

CHAPTER 33

PETERSON IN A STREETCAR

IN CHICAGO

.

In 1916 Joseph Peterson, an eminent psychologist, published a paper on "Illusions of Direction Orientation." It is a goldmine of information, containing detailed observations made by a man who was not only a prominent scientist but also the possessor of a first-rate sense of direction.

But since he, like me, relied heavily on introspection, his paper did not get much attention from his peers, most of whom had never experienced anything of the kind. In fact he complains that most of his fellow faculty failed to understand when he read them his paper. He suggests that the reason for their lack of understanding of his misorienteering experiences was that they had grown up in cities where they could manage without a sense of orientation because they relied on landmarks for their way-finding, whereas he grew up on a farm, using the sun and the stars as cues for direction during cross-country hunting trips.

His first misorienteering experience happened in Chicago in the summer of 1914. It was already dark when he with some friends transferred from a North Side to a South Side streetcar. (Apparently there were two streetcar lines,

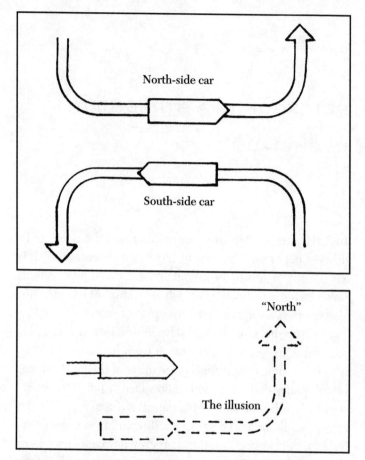

FIGURE 37, TOP. *The probable layout of the transfer point.*
FIGURE 37, BOTTOM. *If Peterson's direction frame was locked in to the interior of the North Side car facing east, and he carried that "lock-in" with him to the South Side car, this would give him the illusion that the South Side car was also facing east. Later, when that car had started up and turned left, he would therefore be under the illusion it was running towards the north.*

one serving the North Side and the other the South Side of Chicago.) He was still standing and giving the transfer tickets to the conductor when the car started and turned to the left. This turn surprised him, since he felt the car was going north afterwards. He asked and was reassured that they were indeed going south, but this did not change his feeling that he was going north. Nor did it put his bearings straight that he caught glimpses of familiar buildings in the dark, or that he recognized cross streets that he felt were going west but knew were going east.

His assumption that the reversal of his sense of direction happened while he was talking to the conductor makes sense, since there was no reason for it to happen before he boarded the South Side car, and it must have been before the car turned. My explanation is this: The North Side car was facing east when it stopped, and the South Side car that he transferred to was facing west (Figure 37). When he was distracted by dealing with the conductor, his spatial system "slipped back" into the orientation of the North Side car, giving him the feeling that the South Side car was also facing east, and therefore went north after turning left.

Peterson goes on to say:

Immediately, however, on stepping out of the street-car in 58th Street and Cottage Grove, in the midst of numerous perfectly familiar objects and buildings I felt an "unwinding sensation" in the head, a sort of vertigo, and presto! the illusion was gone. I saw the car start its motion and continue in the direction it had been going—but now it was due south. . . .

... It is a striking bit of evidence that we are not passive in our perceptions and interpretations, but that everything must be seen in some system of generally consistent relations. Coming with such an erroneous sense of direction into familiar surroundings, one experiences for a time a conflict of two different systems. Sometimes the change is gradual; occasionally, when, as in the present case, the conditions are favorable, it is sudden, with possibly a feeling of being forced by general surrounding objects and circumstances.[56]

What happened was that his reversed direction frame, locked in to the car going north (as he felt it), was forced to switch around by all the familiar landmarks outside the streetcar. This "spatial vertigo," as I call it, he felt as an "unwinding sensation" in the head. Afterwards he was oriented again.

This reminds me of a story told by a teacher who had been transferred from her old school building to an identical new building on the opposite side of the street. She complained that she was always turned around in the new building, feeling that north was south. Apparently her direction frame locked in to the familiar (but turned around) layout of the new building, which caused her sense of direction to reverse. Since this reversal was reinforced every time she returned to the building, it must have stayed with her for as long as she worked there, and even long afterwards if she ever returned to the building.

CHAPTER 34

PROFESSOR PETERSON'S

MISERY IN MINNEAPOLIS

The previous chapter dealt with Joseph Peterson's turn-around in a streetcar in Chicago. In the same article he describes a second "illusion of direction" that he found really troublesome because he could not get rid of it.

In 1915 he secured a teaching position at the University of Minnesota in Minneapolis and moved there with his family. He arrived in the morning by a night train from Omaha. The weather was cloudy when he arrived, and it was several days before the sun came out. When it did, he realized something was wrong with his sense of direction; he felt that the sun was in the east at noon. As with my mis-orientation in Cologne, there was nothing he could do to correct the situation, and unlike me in Cologne, he had to stay there.

Since psychology was his field, he tried to figure out what had caused his problem and came up with three possible causes for his turn-around:

(1) The relaxation when asleep in a moving train, where bodily attitudes could have little chance of functioning in

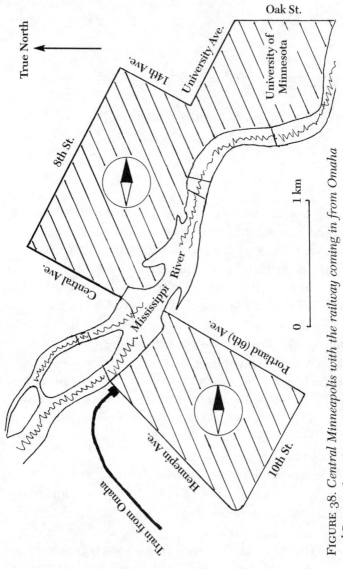

FIGURE 38. *Central Minneapolis with the railway coming in from Omaha and Peterson's misorientation areas.*

the maintenance of orientation even if one were on the alert; (2) the formation of rather permanent associations of parts of the city with the erroneous direction attitude before it could be corrected by noting the position of the sun; (3) the unusual direction of the streets in the section of the city with which these associations were formed. I had become so accustomed to streets following the cardinal points of direction that this one condition had become a fruitful means of orientation when the sun was not visible. One does not usually get the concurrence of so many factors favoring the illusion.[57]

He is certainly right about (2), that he had time to make a misoriented cognitive map of a large area before the sun came out. But I doubt very much that his sleeping in the train had anything to do with it. Nor could it be the unusual street pattern (Figure 38), since it would have caused a misorientation of only 45°. (Many New Yorkers, for example, have their cognitive map of Manhattan misoriented because they feel that the main avenues run north to south when in reality it is more like north-northeast to south-southwest.)[58] Instead I am sure that he was the "victim" of the 90° turn in the railway line just before it enters the station (Figure 38)—just as I was in Cologne! It is not easy to notice a train turning, since one can only see to the side, and he was probably distracted by getting ready to leave the train.

When he got settled and explored the central area of the city, he automatically encoded a cognitive map with north in the east, a map that could not be changed. His excellent way-finding ability that normally took such good care of him as he moved around in his environment was now working against him. Not that he didn't try to get his bearings straight again. He bought a pocket compass and approached what he called the "dangerous" section from various directions, but every time he crossed the invisible borderline into the area, the landmarks in his misoriented cognitive map forced his direction frame to turn 90° clockwise and he was miserable again, feeling that north was in the east!

He noticed, like I did in Paris, that when leaving the misorientation area there was only a gradual change into correct orientation. "At certain places in the borders of these sections there was much uncertainty experienced as to just what really did seem north!"[59]

His wife did not have the same problem, because either she had noticed the train turning into the station, or she had typical "modern urban" way-finding ability, relying mainly on landmarks. His eleven-year-old son, however, suffered the same misorientation, which shows that it ought to be fairly easy to induce misorientations in experiments artificially, provided of course that the subjects have a good enough orientation ability to start out with.

I am reminded of what Harold Gatty wrote about urban way-finding: "Everything depends on your early experiences in a new place when associations which become basic are being established. If at the outset you wrongly orient yourself then it is practically impossible to adjust."[60]

Towards the end of his paper, Peterson gives some good advice how to prevent these misorientations from happening.

If the sun is not visible the true north should be determined as accurately as possible. For this purpose it is a good thing to have, when traveling, a pocket compass. . . . When the sun becomes visible the exploration of the city should begin from some point where the streets follow the cardinal points of the compass, . . . The general rule would be to get acquainted with a strange place only when the usual stimuli by which we determine directions in practical life are most obvious and helpful, and to look about as little as possible at other times . . . Everything depends on the early experiences in a new place, when associations which become basic are being established.[61]

For those with good spatial ability, Peterson's advice is a good idea to follow when moving to a new area. Exaggerated as his admonitions seem to us, they show how unhappy he must have been about his situation in Minneapolis, as does the fact that he left the university three years later, despite having just advanced to be chairman of his department. It was the only way for him to get out of his misery!

It is strange that Peterson does not mention M. Portier's story from Le Croisic in Binet's paper (which he must have read, since he included it in his bibliography), for it is in essence an exact copy of his own experience. There is the same arrival in the early morning after sleeping on the train; the same curve in the railway line just before the station that caused the misorientation; the same period of encoding a misoriented cognitive map of the place before he realized that he had his bearings wrong (not from observing the sun, like Peterson, but from seeing a boat with a known destination leave the harbor in what he felt was the wrong direction); the same fruitless attempt to correct his misorientation with a compass; the same misery: "This annoyance, this indefinable distress, slightly comparable to that which one suffers when dizzy, persisted during the whole time of my stay at Le Croisic (about a month and a half)."[62] And, like Peterson, this unpleasant experience made him take care to orient himself properly afterwards when he arrived in a new place.

What would have happened if Peterson had returned to Minneapolis many years later? The answer is given in the following story.

Virginia went to Berkeley in 1947 to study at the University of California, Berkeley. She took a bus from Los Angeles and after a twelve-hour journey, sleeping on the bus, she arrived at the bus station in Berkeley on a cloudy morning. She took a taxi from the bus station to the dormitory. Going uphill on Bancroft Way in the taxi, she felt she was going north, when in fact she was going almost due east.

She noticed her misorientation later when she felt the sun rise in the north and set in the south. During the eight years she lived in Berkeley the misorientation remained unchanged. When finding her way around in Berkeley, she therefore had to disregard her *feeling* for directions and instead consciously orient herself by her *knowledge* that San Francisco Bay is west in the area. It is remarkable that when Virginia returned to Berkeley for a visit after thirty years' absence, the misorientation was still the same. We can therefore be sure that if Peterson had returned to

Minneapolis, even many years afterwards, he would have experienced the same misorientation as before.

It is a common misconception among those who have become misoriented somewhere, that if they had been able to return to the area more often, the misorientation would have disappeared. If that were true, then Virginia, who returned to the area where she was misoriented practically every day for eight years, ought to have erased her misorientation. This is not how the spatial system works. In Virginia's case it was this daily revival of the misoriented cognitive map that, rather than erasing it, engraved it so deeply that it was still impossible to correct when she returned thirty years later.

Passini, whose excellent spatial system took him straight back to the car after several hours of picking mushrooms in a forest, reports the following: "During my first visit to Manhattan, I imagined the island upside down. What is south in reality, the financial district, appears as being north in my map, Even today, while I write this text, my map is not properly organized. That first image has to be rotated by 180° to get it in place."[63]

He does not mention how he arrived in Manhattan the first time, but a good guess is that the last part of the journey was underground, as it is at Grand Central Terminal or by most subways. He then happened to get his sense of direction turned around when he emerged into the street, as I did in Paris.

The moral of these stories is that a strong sense of direction, which is such an asset in the wilderness, sometimes gets turned around in the artificial urban environment and becomes a nuisance, since it cannot be corrected.

A French professor once told me that he had a sense of direction that did not give him any problem. For him directions were the opposite of absolute. He would sit at the table and talk about a place and gesture towards it but in the wrong direction. When others pointed out his mistake, he would just say: "If I had been sitting on the other side of the table, I would have been right!" In the Paris métro, he had no idea whatsoever about directions. He just oriented himself from the map or from experience when he came out in the open. He's obviously a very well adjusted modern man. Living in a metropolis with underground transportation one is probably *better off* with a weak sense of direction, since a strong one would easily go wrong under those circumstances, and be harder, or even impossible, to correct.

As I write this chapter a faded memory emerges about how a Chinese emperor in antiquity had solved the misorientation problem. When traveling, he surrounded himself with a living compass rose: four standard bearers with banners indicating the cardinal directions. These "cardinal directors" had to see to it that regardless of where the emperor went, he would always see them in the proper directions. Thus, as long as they performed their task correctly, the emperor could never become misoriented.[64] Those of us with more modest resources might find it easier to follow Professor Peterson's advice and get a pocket compass.

CHAPTER 35

TALES OF

A COSMOPOLITAN LADY

By accident (as usual) I ran into an interesting paper from 1927, written as a story in the first person singular: "I did this and I did that and then this strange thing happened to me, etc." Back then, articles in scientific journals could still have charm; the author was allowed to be delightfully human, somebody who through his or her writing could make friends with the readers! Now, they are all long since forgotten, hidden away in the scientific attic, never mentioned, never referred to, until somebody like me, poking around haphazardly, stumbles on them and resurrects them from oblivion, wakes them up from their Sleeping Beauty trance.

The article in question was "Die Orientierungstäuschungen" (The Illusions of Orientation) by Franziska Baumgarten. She describes several misorientations that happened to her when living and staying in various European cities. I will concentrate on two stories that I find particularly interesting.

She twice visited Warsaw. The first time she walked south from the Wiener station to her lodgings, and made a correct cognitive map of the area with which she became familiar. The second time, a few years later, she took a taxi from the Wiener station to her apartment, which was a little farther south than the one she lived in the first time. This time she made a reversed cognitive map of the same area that she was familiar with during the first visit. She tried repeatedly to correct her orientation during the two weeks she stayed there, but to no avail. In fact, when writing her paper long afterwards, this reversed map was still vivid in her memory.

The most likely reason for her reversal of orientation during her second visit is that, instead of walking from the station, she took a taxi to her apartment. It is always easier to get misoriented when being passively transported in a car than when actively walking. She was therefore already turned around when she arrived at her apartment, and this reversal stayed with her when she later walked around and encoded a cognitive map of the area.

What is surprising is that she managed to turn around the area she was already familiar with from her first visit and make a new reversed map of it. This is usually impossible, at least for persons with a good sense of direction. The only explanation I can think of is that her map from the first visit had faded so much that she was able, not to turn it around, but rather to replace it with a completely new one. I can compare this with what happened to me when I revisited the reversal area by the Château d'Eau métro station in Paris some forty years after the first visit. Then my old reversed map was so faded that I was able to encode a new correct cognitive map.

She visited Paris several times, and once she had a most remarkable experience. She arrived at night at a hotel in Boulevard St. Michel near Place de la Sorbonne (Figure 39), an area she knew very well.

> The next morning I went to a café in the street, a few steps to the left of the hotel; on leaving the café I wanted to go towards Notre-Dame and firmly believing I had taken the right way, I went up the street. When I got to the Luxembourg Gardens I noticed my mistake, but suddenly I could no longer orient myself—how did I get here and where was my hotel? Everything was as if turned around. On my way back, I came again to Place de la Sorbonne, *which I did not recognize at all*, and I could not get rid of my disorientation, which moreover gave me a very unpleasant feeling of uncertainty, until I sat down and made a drawing of the earlier so familiar street and put in the location of the present hotel.[65]

A very interesting story. She arrived at night probably by taxi or possibly the subway. This is the ideal situation for getting misoriented, especially since she had not bothered to picture where the hotel lay in the boulevard (that is to say, included it correctly in her old cognitive map of the area).

Her explanation that she automatically continued to the left from the café the next morning because it lay to the left of the hotel, and that this happened to be the wrong direction and therefore *caused* the misorientation, does not make sense. It was the other way around. It was her misorientation (already established when she arrived at the

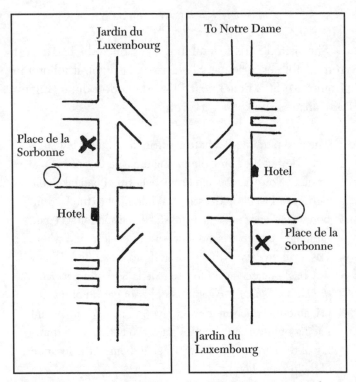

FIGURE 39, LEFT. *Map sketch made from her new reversed cognitive map of Boulevard St. Michel.*

FIGURE 39, RIGHT. *Map sketch made from her old oriented cognitive map of the boulevard. When she is following her new reversed map and passes a familiar landmark X, she cannot recognize it, because in her old cognitive map it is on the opposite side of the boulevard.*

hotel the evening before) that made her automatically take off to the left, because that is where she saw Notre-Dame in her misoriented cognitive map. Thus, instead of the reversal being caused by her setting out in the wrong direction from her hotel, it was her already-present reversal that caused her to take off in the wrong direction (which of course she saw as the correct direction in her cognitive map at that moment).

When she tried to find an explanation for the reversal, she went back in her analysis from the Luxembourg Gardens to where she started the walk, and therefore naturally decided that the mistake was made accidentally when she left the café. And one cannot blame her, since she did not know how her spatial system worked, nor did anybody else at that time.

During her walk that morning one could say that she passed through an environment that existed only in her mind, not in the real world. She was under the illusion that she was going towards Notre Dame, when in fact she was going in the opposite direction towards the Luxembourg Gardens. As long as what she saw along the street did not seriously contradict that illusion, she could live happily with it. In fact she passed Place de la Sorbonne soon after leaving the café and failed to recognize it, nor did she recognize it on her way back, in spite of knowing at that time that her directions must be reversed.

It seems strange that she could walk in the wrong direction along a familiar street (she had lived in the area for a whole year not too long ago) without noticing her mistake. Ironically, it was to a large extent the fact that she knew the area so well that misled her. Being confident that she was on the right track, she trusted her unconscious way-finding system to guide her. Instead of looking for familiar landmarks along the route to show her where she was, she relied on her dead reckoning system to tell her how far she had proceeded.

Note also that she felt that north was south during her walk. This prevented her from recognizing familiar landmarks, since they all appeared turned around and on the wrong side of the street.

CHAPTER 36

THE TOPSY-TURVY

GLOBE-TROTTER

Of all the many strange stories about reversals of orientation, this one beats them all. For A. Kirschmann, a German born around the middle of the nineteenth century, not just a limited area but the whole world was reversed, except for the area around his childhood home and some small "islands" with correct orientation. In the seventy-five years since the publication, nobody has stepped up to the plate to explain this!

Kirschmann starts by describing a childhood experience. He was about five years old when he went to see a play in the Philipp Kleinschen Auditorium in his hometown of Oberstein an der Nahe. There he felt that he faced south when he looked at the stage, when in fact he was facing north. When he went there later, even if he tried to keep track of all the turns on the way up to the auditorium, he still had the same illusion.

Once his spatial system had made a cognitive map of the auditorium with the directions included, the map could not be corrected. This is true for everybody with good spatial ability, and—since he states that this kind of illusion had not

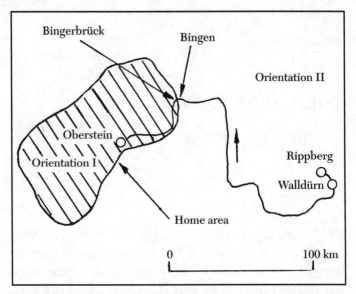

FIGURE 40. *The return journey from Rippberg to Oberstein. After being in orientation II from Rippberg to Bingen, he got turned around back into orientation I when he entered his home area at Bingerbrück.*

happened anywhere else in his home area—it shows that he must indeed have had good spatial ability. He goes on:

> In 1878 I was private tutor in the family of a factory owner in Rippberg by Walldürn in Baden. After a few weeks there it struck me . . . that there the sun, compared with the orientation in my home area, rose in the west and set in the east. I was therefore not oriented in Rippberg as I was in my home area, but as in the hall of my hometown, the Ph. Kleinschen Auditorium. This was also the case during the many walks I took with my pupils over the entire region; the same orientation continued in all the Odenwald area, in Walldürn, in Amorbach, Miltenberg etc.
>
> When I went home for the first time over the holidays, I established that I was still misoriented in Mainz, in Bingen; even during the journey from Bingen to Bingerbrück I looked for the mouth of the Nahe on the wrong side. However, just before Bingerbrück things suddenly turned around, and I was again in my home area orientation.[66]

He calls his home area "orientation I," and outside "orientation II" (Figure 40).

He is not only in "orientation II" just around Rippberg; whenever he goes outside his home area he stays in "orientation II." And he travels far and wide, all the way to North and South America both on the Atlantic and the Pacific side.

Only in a few small places outside his home area is he in "orientation I." However, as I see it, this is not because he managed to get his bearings straight in these locations, but rather because he got turned around 180° again and thus ended up again in "orientation I."

I have never heard or read about anybody else who suffered from this kind of extended reversal in time and space.

He attempts to explain what happened by proposing that space has a double nature. He then expresses his belief that practically all people can experience this reversal of orientation illusion and that "this illusion fundamentally is not an illusion, but only the result of the described double nature of our space,"[67] which must have made him feel good. It is always nice to know that one has company in one's misery, but it shows what can happen when one generalizes out of a sample of one, especially when that one is oneself. He was mistaken of course, and we can all be grateful for that!

Let us now see if we can come up with a more down-to-earth explanation for his spatial problem. It is clear from his story that during the first part of his childhood and youth his orientation was generally correct. He therefore had a correct cognitive map of his home area, between the Rhine, Saar and Mosel rivers (orientation I in Figure 40). After his move to Rippberg in Baden and for the rest of his life outside his home area, his orientation was generally reversed, resulting in a reversed cognitive map of the rest of the world (orientation II). Thus the key to solving the problem is to find out what happened inside his head when he moved to Rippberg.

How his direction frame got reversed during this journey we can only guess. He does not give any details in his article as to how he traveled to Rippberg, but since it was before the days of the automobile, we can assume he must have come by train. Rippberg is east of Oberstein, and the railway through Rippberg runs approximately east–west, but there is the possibility that he might have traveled via Walldürn. The fact that he described Rippberg as being "by Walldürn" indicates that Walldürn (to the east of Rippberg) was the central town in the area at the time. In that case, traveling from Oberstein to Rippberg via Walldürn, he would have arrived in Rippberg from the east, "through the back door," so to speak, which could explain the reversal, especially if he came at night or on a cloudy day.

However it happened, his direction frame must have been reversed when he arrived in Rippberg, before he started exploring the area. This reversal caused him to make a cognitive map of the village and the surroundings with north, for all purposes, south. This is similar to Peterson's Minneapolis experience, but Peterson realized his mistake as soon as the sun came out, which limited his reversal area to the streets he had explored during his initial cloudy days in the city, whereas Kirschmann went on enlarging his cognitive map even after he noticed that the sun "rose in the west and set in the east" until the map included the rest of Germany outside his home area, the countries of Europe he had visited, the Atlantic, the Americas and the Pacific Ocean—in short, the rest of the world he covered during his extended journeys! How could his spatial system do that to him?

One statement in his story gives a clue: it took him *a few weeks* before he realized *what he had not paid attention to until then*: that the sun rose in the west and set in the east, when he compared its movements with those in his home area. During these "few weeks" his spatial system had had time to make a reversed cognitive map, not only of the Rippberg terrain, but also of the sun in the sky above it.

Just as outdoor men have a "cognitive sun compass" that automatically keeps their bearings straight when exploring new areas on sunny days, the unlucky Kirschmann developed a "reversed cognitive sun compass" with north appearing south. By the time he noticed that something was wrong, this compass was already well established and could not be corrected. It therefore stayed reversed and became even more entrenched during the rest of the year he stayed in Rippberg.

What happened to Kirschmann in Rippberg was a total reprogramming of his direction frame. In order to understand it, we must look at how his original programming took place in Oberstein an der Nahe, which is fairly easy to do, since all of us who have a good spatial system go through basically the same process.

As toddlers we make a cognitive map of the inside of our home, and this map is then expanded as we get old enough to go out and explore the surrounding area. At this early stage, the direction frame is held in place by landmarks encoded in our cognitive map. The points in our compass rose are simply directions towards landmarks, such as "up the street towards the church," or "down the street towards the market square."

Later when we explore further, especially if we leave the streets and go cross-country, we will include the sun in its movement across the sky as a cue for direction, holding the direction frame oriented, and we will start using direction labels like north, south, east, and west. We have thus developed a direction frame that is not dependent on known landmarks in a familiar area but that functions anywhere—provided we stay on roughly the same latitude, I should add—since it relies on a celestial "landmark."

The reprogramming of the direction frame as an adult that Kirschmann went through must have followed a similar course but within a shorter time frame. Thus, as a result of what must have been a rare series of unfortunate circumstances, he ended up with a topsy-turvy spatial system that stayed with him for the rest of his life.

I must confess that I could never imagine such a directional disaster happening to anybody. It makes me recall how I felt when I realized I was "turned around" in Cologne, how I came down to the Rhine and saw the water flowing south. It made me worry that all Germany was going to be "turned around" afterwards, causing feelings of anxiety bordering on panic.

Since I have only found one example in the literature and have never heard anybody complain about it, I have assumed that it is a very rare problem. But it might not be. Those who suffer from it might not realize it is something unusual, and if they do, they might be too embarrassed to talk about it.

Would it have been possible for Kirschmann to get rid of his problem? The normal cure is to stay away from the reversal area. This is easy when one only has a few small "reversal islands" to avoid, but in Kirschmann's case it would have meant returning to his home area, where his direction frame, encoded in childhood, was correct—and staying there for the rest of his life. There would still have been a conflict with his reversed cognitive sun compass, but the terrestrial landmarks would have prevented the sun from triggering a reversal of his orientation. I say this considering what happened to me in Cologne when I saw the sun rising in what I felt was the west. The misoriented terrestrial landmarks in my (only a few minutes old) cognitive map were enough to prevent the sun from reversing the map.

It is likely that if he had stayed long enough in his home area, his cognitive sun compass would have corrected itself—after all, it reversed itself completely and permanently during the year he lived in Rippberg. He would then have been able to explore *new* areas outside his home area and make properly oriented cognitive maps of them. However, if he ever went to a familiar area outside his home area, an area where he already had a reversed cognitive map, then the landmarks in this map he would have caused his direction frame to spin around and he would have been back in his misery, like I was when I returned with the steamer to Cologne. Only if he had waited so long that the old reversed cognitive map had had time to fade could he have made a new properly oriented one, like I did when I returned to my reversal area in Paris some forty years afterwards.

One is tempted to conclude that when somebody can be "turned around" all over the world, it shows that humans cannot have a magnetic sense. But that is not necessarily so. For example, we now know that pigeons have a magnetic sense and use it as the foundation upon which to build their sun compass and as a default compass in cloudy weather. But even in pigeons the sense is so subtle that it took many negative experiments before it was positively shown to exist.[68]

Since in Rippberg the whole direction frame was reprogrammed, changing by 180° the strong hold on it by the sun, it is easy to imagine that the weak hold by a much more subtle magnetic sense also could have been changed 180° during the year in Rippberg.

Finally, we must not let Kirschmann's far-out explanation about the double nature of space detract from the value of his article. His paper is unique and without it we would never have known that this remarkable illusion of orientation could happen.

But it would be nice to get more than one such story and thus be able to make comparisons. Therefore, if any of you readers are living with a similar reversal of your direction frame, i.e., one that always leads to a reversed cognitive map when you are exploring a new area, please, PLEASE, *PLEASE* tell me about it! Readers may write to me c/o Scribner, 1230 Ave. of the Americas, New York, N.Y. 10020.

CHAPTER 37

CAUSES OF MISORIENTATION

We can safely assume that evolution has not prepared us for being passively transported. Until only a few thousand years ago—a totally insignificant passage of time in evolution—all our movement occurred by means of our own legs; to get to a destination we had to walk there. No wonder we get into trouble when we are displaced too far and too fast. The spatial system, unable to keep up, is apt to simplify the cognitive map it encodes of the route. Curved roads are straightened out and oblique intersections are neatened to right angles.

This is especially true if the sun is hidden in clouds and therefore cannot serve as a cue for directions, and when the visibility is reduced, for example in misty weather and at night.

The kind of vehicle also plays a role. A motorcyclist is in a better position than a driver of a car, since his field of vision is unobstructed in all directions and he can see the sun even when it is high in the sky. The driver of a car in turn is better off than a backseat passenger. A traveler in a train is unable to see ahead and therefore cannot evaluate to what degree a train turns in a curve. I will describe a couple of real-world experiences to illustrate this.

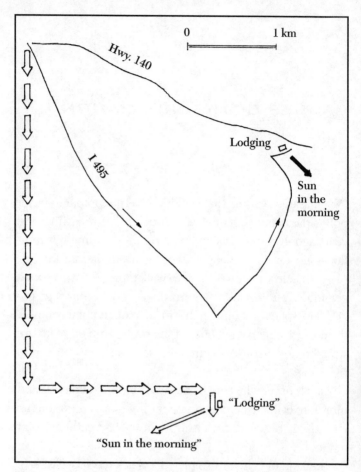

FIGURE 41. *The route followed by my daughter. The white arrows show where she felt she was going. With her sense of direction turned 110° she felt the sun was in the WSW when it was actually in the SE in the morning.*

My daughter was a passenger in a car from the Boston airport west on the Massachusetts Turnpike, then south on Interstate 495 to the exit nearest the town of Franklin, and finally onto Cross St., where she was going to stay. When the sun came out in the morning two days later, she was surprised not to see it in the southeast as she expected but in what she felt was west-southwest. Looking at a map she found the cause. She had assumed that Interstate 495 proceeded straight south all the way and that the succession of streets that followed were straight and all the intersections at right angles (Figure 41).

Sometimes urban planners unintentionally—I hope!—make streets that can turn around the direction frame of unsuspecting first-time visitors. A beautiful example of this is Reagan Road in Mira Mesa, a San Diego suburb. It runs in a half circle with a radius of 550 meters (Figure 42). My friend Lorna came to grief here on a rainy and misty day as a passenger in a car. She failed to notice the slow turning of the road and therefore came up with a mental picture of Reagan Road running straight south and the western intersection with Mira Mesa Boulevard turned around. As a result, when she went there afterwards she had to do everything backwards. When she wanted to go north, she had to go towards what she felt was south.

A special case of this I call the Buckman Springs effect, since many of those who travel from San Diego east over the mountains on Interstate 8, myself included, experience it in the Buckman Springs Safety Rest Area. Knowing that the freeway runs generally east, we assume it runs east also at Buckman Springs, but in fact it runs nearly south

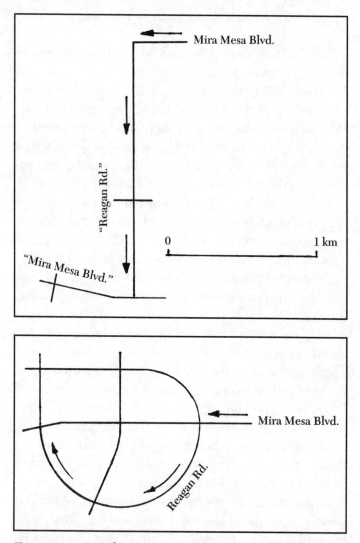

FIGURE 42, TOP. *The route Lorna felt she was following.*
FIGURE 42, BOTTOM. *The topographic map of her route.*

there (Figure 43). Hence, at that location our direction frame is almost 90° off; what we feel is north is in fact nearly east. Normally, most people passing through are not even aware of it. However, when we stop at such a location and try to orient ourselves, we will, if our direction frame is strong enough, experience a conflict between the conscious knowledge of where north is and the unconscious feeling that it is somewhere else.

Since the safety rest area at Buckman Springs is an ideal location for interviews about the sense of direction, I made a little experiment there one morning, asking subjects to point to the north, in order to see what percentage would get fooled by the freeway running north–south instead of east–west.

The weather was cloudy with only a few thinner patches where the sun occasionally showed (but care was taken *not* to ask anybody to point when the sun could be discerned). Twenty-seven persons were interviewed. Of them fifteen pointed roughly to the "false north" (bearing 73°) perpendicular to the freeway, eleven pointed roughly to true north, and only one pointed in between to bearing 40°.

As was to be expected with the sun invisible, the majority of the subjects went by the direction of the freeway when orienting themselves. Actually, I was surprised that not more of them did. The reason for this may be that many of those who were interviewed had been in the rest area previously, when the sun was shining (the normal weather in Southern California), and established their orientation at that time. And once a cognitive map has been encoded of an area, the directions in it will not change.

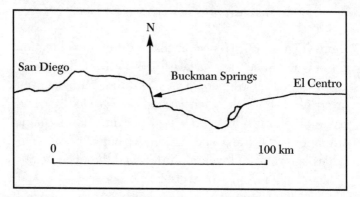

FIGURE 43. *Interstate 8 from San Diego to El Centro runs generally east, but at Buckman Springs it runs NNW–SSE.*

The "Buckman Springs effect" is not limited to roads—any linear feature can cause this confusion. A friend of mine went on a trip along the Colorado River in an area where it runs generally west but still meanders strongly. He stopped at a place where the river flowed south and hiked towards the west, straight away from the river. It was a very strange and unsettling experience for him, since his compass showed he was going west but he felt strongly that he was going north. Even when telling the story to me long after-wards, his confusion was evident and he was unable to give a really coherent description of what had happened. In fact, it was so painful for him to revive this memory that he didn't want to discuss it further when I tried to get more details as to how it had happened.

Influenced by his overall mental picture of the Col-orado River as running westwards, he had assumed that the river was running west in a place where it was running south. Hence, when he set out to walk towards the west using map and compass, he had to fight the unconscious conviction that he was going north instead.

This strategy of imagining a road or river as running always in its overall direction that is used by our spatial sys-tem in order to simplify matters when we are covering long distances in a short time can be seen as a sensible adap-tation to modern conditions. Motorized travel on roads, an innovation of the last century, has resulted in a sharp increase of visual input for the driver when traversing a new area. It is therefore understandable that for some of us, maybe even most of us, our spatial system stops monitoring direction and instead assumes that the road goes in a

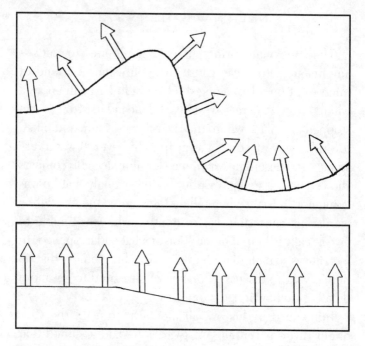

FIGURE 44, TOP. *A topographic map of a road with the arrows indicating one subject's feeling where north is.*

FIGURE 44, BOTTOM. *By aligning the arrows in the top sketch one gets the subject's cognitive map of the road.*

straight line from start to finish. This results in a rather peculiar cognitive map whereby the road is running east all the time if we take Interstate 8 from San Diego over the mountains as an example, but where the curves are still there as landmarks.

An interesting experiment would be to take a busload of subjects for a ride along a winding road and ask them to point to the north at regular intervals. (In order to prevent the subjects from calculating, rather than feeling, where north is, it might be necessary to keep their minds occupied in some way.) It would then be possible to draw a kind of cognitive map of the road for each participant simply by adjusting the direction of the road at each checkpoint so that the estimated north is straight up on the sketch. A likely result of such an experiment is shown in Figure 44.

Although we normally do not register the changes in direction of a road, in some cases it is made more obvious. I recall driving to the desert along I-8 on a November evening with the constellation Orion rising due east as always, and being surprised, as I passed through all the curves over the mountains, to see Orion move sometimes far out to the left and sometimes far out to the right instead of staying straight ahead, as I felt it ought to do.

CHAPTER 38

THE SAN FRANCISCO

EFFECT

These pages were written during a hike in a desert wash with flowering smoke trees, hiking until the body starts to tire, writing in the shade of a smoke tree, hiking again, writing again. . . . A delightful way of getting thoughts down on paper, especially if one does not overdo it. Five hours is enough. The rest of the day, the slowly flowing golden hours of morning and late afternoon, can most profitably be spent in grateful contemplation of being alive surrounded by beautiful, undisturbed nature and majestic mountains. This is when the gremlins in the unconscious can go to work breaking up outgrown theories and then reassembling them together with new facts to form more adequate ones.

Our direction frame has a tendency to rely on outstanding features like the ocean or a mountain range. This can cause problems if the individual moves to an area with a similar feature in a different direction. For example, somebody who grew up on the East Coast and then moved to the West Coast might feel that the ocean was still in the east and

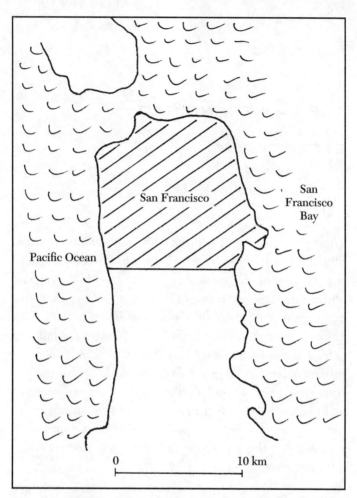

FIGURE 45. *In San Francisco there is salt water not only in the west but also in the north and east. This can confuse Californians used to seeing the ocean in the west.*

hence north is south. This turned-around direction frame can be distressing to live with and difficult to adapt to.

When I was interviewing participants at an orienteering meet, I ran into this effect. "I have trouble with directions in San Francisco," one man said, "too much salt water. When I smell the ocean I know it's west, but that does not work in San Francisco, where there is salt water all over the place." I asked him where he grew up. "San Diego," he said, "lived there all my life." That explained it: His sense of direction was locked in to the ocean. While most of us have a "compass rose" with north, south, east, and west; he had a very special one: ocean (west), inland (east), along the coast to the right (north), and the coast to the left (south). Nice and easy, no problems—as long as he did not go to San Francisco, where there is sea water also to the north and the east (Figure 45).

I recently talked to a university student who described exactly the same condition. She grew up in San Diego and gets confused every time she goes to San Francisco. And she told me that spontaneously, before I had mentioned this "San Francisco effect."

A lady who grew up by the ocean and later moved to the Midwest complained that she had trouble with her orientation in the new area because she missed the ocean, her old mainstay for keeping her bearings straight.

A young woman who grew up in Boston and who now lives in La Jolla, California, suffers from a rather severe case. For her the ocean indicates the east and nothing can change that. "Every time I see a compass here it is pointing the wrong way," she complains (Figure 46). When she thinks of

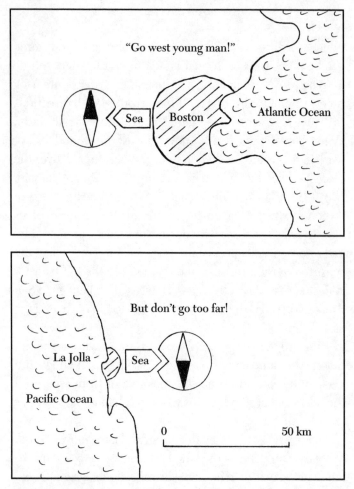

FIGURE 46, TOP. *In Boston where she grew up her sense of direction is correct.*

FIGURE 46, BOTTOM. *In La Jolla, with her direction frame locked in to the ocean in what she feels is east, her sense of direction is reversed.*

Canada, she puts it to the left along the coast, where Mexico is. And she gets no understanding from her husband, who grew up in the Midwest, and so did not develop a direction link to an ocean. She told me that she plans to return to live in Boston soon. Probably a smart move: the distress caused by the conflict between her unconscious conviction and reality could very well prove harmful to her health in the long run. Poor Professor Peterson, in a similar situation in Minneapolis, also had to move.

A friend of mine told me the following story, which shows how even a mountain range can serve as mainstay for the direction frame.

> I lived in Western Colorado (Grand Junction) from ages 4 to 10. The mountains were to the east. We moved to Pasadena when I was 10. The mountains there are to the north. I lived there until I was in my twenties and it always seemed to me that the mountains were east. In other words, *all* mountains are to the east, was my orientation.
>
> I still have that problem (at 60 plus years) when I'm in the Eastern Sierra, although to a much lesser extent, it seems that the mountains *should* be east even though they're west.

We have here several examples of how a dominant feature in the surroundings becomes so firmly linked to the direction frame during childhood that it actually becomes its main support. This is an advantage only as long as we stay in the same area. However, as soon as we move to an area where a similar dominating feature (ocean, mountain range) is in a different direction, we will have problems. The direction frame that normally takes good care of us when we move around will work against us; in order to get to the right place we have to move in what we *feel* is the wrong direction. And the irony is that the stronger our sense of direction, the more severe the problems. Somebody with a weak direction frame will have no problems at all, or rather the problems caused by lacking the support of a strong direction frame will be the same regardless of the whereabouts.

One might wonder how a handicap like this survived in evolution. The answer, as I see it, is that humans are designed to function in a relatively limited area. As long as we stay there, our direction frame will take good care of us; it is only since we became globe-trotters that this feature of our direction frame has become a liability.

It should be noted here that children in nomadic tribes moving long distances seasonally never stay long enough in the same area for their direction frames to have time to lock in to a particular terrain feature. They will therefore have to rely on other supports, like the sun, to stay oriented. Later in life they will consequently be able to orient themselves well regardless of where they are, as long as they do not move to a different latitude where the sun path is different. A nomadic lifestyle should thus give us no problems, provided we follow it from birth.

It is obvious that a person with a strong direction frame forced by circumstances to live in an area where it is reversed, like the woman from Boston living in La Jolla, will feel ill-at-ease.

The man whose direction frame was locked in to a

mountain range stated that he still has that problem, *although to a much lesser extent,* now that he is getting older. This points to the likelihood that the strength of the direction frame diminishes with age. (At age seventy-five mine definitely has!) At least, those sufferers living in an area where their direction frame is out of line and are unable to move somewhere else have the consolation that the situation is likely to improve later in life.

I am reminded of a passage at the beginning of Plato's *Republic* where Sophocles, when asked if he could still possess a woman in his old age, answered: "My friend, it is with the greatest satisfaction that I have fled it, as if set free from a violent-tempered and savage master." For him, this was also an improvement.

Those who grow up in an area lacking dominant terrain features never have a chance to develop the direction frame fixation described. They can therefore adapt more easily to any location later in life without problems. Apparently childhood experiences have a great influence on the development of the sense of orientation. The more spatially challenging the childhood environment, the better this sense ought to turn out.

As my old Swedish friend proudly put it: *Forest people don't get lost!*

CHAPTER 39

CROSSING A RIDGE
WITHOUT GETTING
TO THE OTHER SIDE

I will now describe what can happen when one goes cross-country through a generally familiar area following an overall cognitive map. This is usually no problem: one knows the starting point and the direction to follow to the destination, and has a rough idea of what lies in between. I have done it myself successfully many times. However, under exceptional circumstances, strange things can happen, as the following story shows.

Professor Bleckmann, a German zoologist, told me in a letter that as a young man he was out hiking in the forest near the city of Siegen. After reaching a small road from Siegen to Eisern he decided to go cross-country over a ridge to the other side, where a major highway would take him back to Siegen. He knew the ridge well, since he had hiked a trail running along the crest several times.

Like every experienced hiker, he used the slope as a guide for direction through the dense forest, going straight uphill until he reached the crest. There he was surprised not to find the trail. He decided that he must somehow have crossed it without noticing it, and continued down the other side, still guided by the slope. After a while he reached a curved, narrow road that he could not identify, and did not at all expect in this part of the forest.

This astonished him. It had such an impact that he still had a vivid picture of it when writing the letter thirty years later. He followed this road to the right—what he erroneously felt was the direction to Siegen. After a while he realized he was on the small road from Siegen to Eisern and later found the place where he had crossed it before going up the ridge. Thus he must have walked in a full circle! He could hardly believe that such a thing could happen.

As a compulsive problem solver, I was delighted to have such a challenging mystery to tackle. I read this letter several times and tried to figure out what could have happened, but to no avail. How could he cross a ridge and come down *on the same side*?

Finally I decided the problem was unsolvable and put it out of my mind. But my unconscious mind must have taken over the investigation, for some hours later, when occupied with something completely unrelated, it dawned on me. Instead of crossing the ridge, he must have crossed a lateral spur going out from the main ridge towards the valley he came from. By going uphill all the time, he would then automatically have followed a curved route perpendicular to the contour lines, and ended up on the summit of the spur thinking it was the main ridge. In the dense forest it was impossible for him to notice that he had deviated. Thus, the direction frame in his spatial system had turned

90° by the time he reached the summit. Therefore, in his cognitive map made during the climb, the spur he was on fitted exactly—both in location and direction—the main ridge in his overall cognitive map of the area, except for one astonishing detail—the trail along the ridge was missing.

Then he crossed the spur and walked down the slope on the other side. Going straight downhill in the dense forest, he again deviated as the slope changed direction; hence when he reached the narrow road from Siegen to Eisern his direction frame was 180° off (Figure 47). In other words he was completely "turned around," feeling that north was south and east was west. He therefore turned right, towards Siegen as he felt it, but in reality away from the city.

I sent Professor Bleckmann a letter with my hypothesis about what had happened, together with an old, detailed topographic map of the area that I had managed to dig out of the university library. In his reply he agreed with my explanation and marked on the map the ridge he had tried to cross and two lateral spurs, each of which could have led him astray (Figure 48).

The critical reader might argue here: Why did he not turn left when he reached the top of the spur and failed to find the trail that he knew had to be on the ridge? It seems the obvious thing to do when looking at the topographic map of the area. The answer is that he was *not* looking at a topographic map, he was looking at his cognitive map and this map was turned 90°; what he felt was north at that point was actually east. The way his cognitive map was misoriented, if he went left, he would have felt he was going south *along* the ridge and *away* from Siegen, the place he

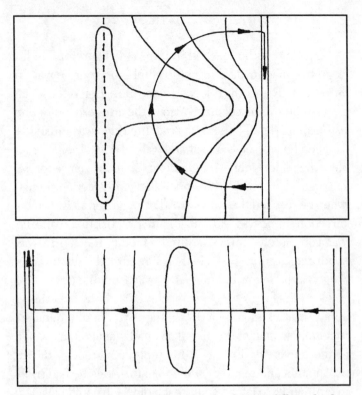

FIGURE 47, TOP. *Topographic sketch of route followed when going straight uphill and downhill across a spur.*

FIGURE 47, BOTTOM. *Resulting cognitive map of the route followed.*

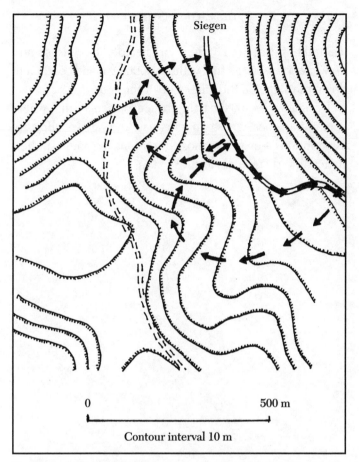

FIGURE 48. *Two possible routes for the turnaround. The northern one seems the most likely.*

wanted to return to. He therefore decided he must some-how have crossed the trail without noticing it and continued down the slope on the other side of the spur.

I can assure you, speaking from experience, that when something goes wrong in our way-finding we cling to assumptions—even the most unreasonable ones—rather than admit that we are lost.

CHAPTER 40

DETERIORATION

OF OUR SPATIAL SYSTEM

IN OLD AGE

Lately, I have experienced several reversals of direction, many of them indoors. I go into the room with my bearings straight and then, when leaving, I turn in the wrong direction as I enter the corridor outside. This had not happened to me before my seventies. It must therefore, at least in my case, be caused by the deterioration of the spatial system in old age.

MISORIENTATION IN MY OWN LIVING ROOM

One evening at eight, when I was tired after a night with very little sleep and a long eventful day, I sat down to watch TV as I usually do, in the northwest corner of our living room, in my usual chair looking north. After a while I noticed to my surprise that I felt I was instead in the southwest corner of the living room facing west. It was as if my

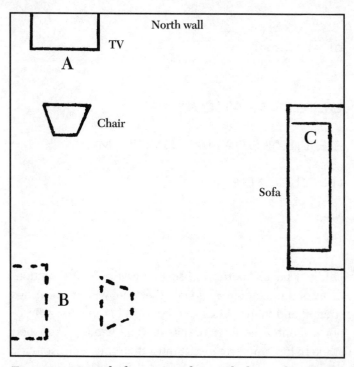

FIGURE 49. *I was looking as usual towards the north at the TV set at A. After a while I was surprised to feel that I was instead at B looking westwards at the TV set. The reason for this misorientation could be that I normally sit looking westwards in the sofa at C when reading, writing, or just relaxing.*

TV corner had moved from A to B in the living room (Figure 49).

One possible explanation: tired as I was, I might have dozed off a little as I was looking at the program and, upon awakening, felt I was looking in the same direction as when I read—and often drift off into sleep—on my favorite corner of the sofa. But in that case it would be remarkable that this feeling of misorientation was not instantly corrected when I looked at the TV set, the most outstanding landmark in this, the most familiar of all the rooms I know. For what other object in a living room catches one's full attention so often and for such long periods of time!

The morning after it happened, this misoriented map of myself in front of the TV set was preserved in my memory. I could clearly see episodes in the program, and I still looked at them in the wrong corner and in the wrong direction, just the way they had been encoded. Thus, not only did it happen, but it was also encoded deeply enough to be retrieved unchanged afterwards.

This shows that something as immaterial as a TV program, remembered several days after it was shown, has not only a point in time attached to it (the time of day fairly exactly, the date more and more approximately as time goes by) but also a very definite location and orientation (the place where I felt the TV set was located and the way it was turned) and that this location in the cognitive map, like everything produced by our spatial system, is very robust: most likely it will stay unchanged—and be remembered that way—as long as I remember this TV program itself.

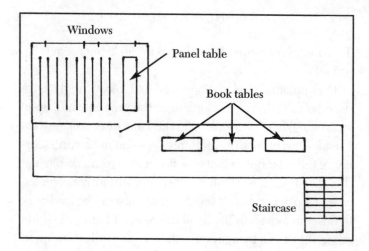

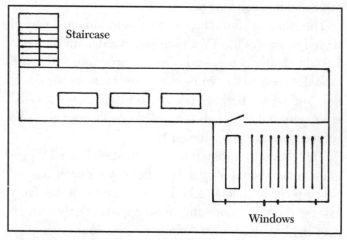

FIGURE 50, TOP. *The correct map of the corridor with the auditorium towards the north. This is how I now, a year later, still visualize the corridor.*

FIGURE 50, BOTTOM. *My reversed map with the auditorium towards the south. This is how I still visualize the auditorium.*

Obviously I have now joined the ranks of Darwin's "old and feeble," capable of becoming so mentally overtired that even unimaginable slips of the spatial system can happen.

REVERSAL IN AN AUDITORIUM IN PASADENA

Another reversal of my sense of direction happened when I was listening to an evening panel discussion in January in an auditorium at Pasadena City College. Leaving the room after the meeting had ended, I was surprised to see that the book tables in the corridor outside had been moved over to the opposite side. I also took off in the wrong direction to get to the staircase down to the first floor. Only when my wife pointed out I was going the wrong way did I realize that my sense of direction was reversed (Figure 50). This turned my cognitive map of the corridor around to the correct orientation. Afterwards the book tables were on the side where they had always been and the staircase was in its proper place. This turnaround of the direction frame took place very smoothly, without any discomfort. There was none of the violent "spatial vertigo" I experienced half a century earlier in Cologne and Paris, when my direction frame was forced to turn around.

As before, tiredness must have played an important role. The turnaround happened in the evening and after a one-hour walk around Pasadena just before the meeting. The tiredness may have been due, as well, to the decrease in stamina that comes with old age.

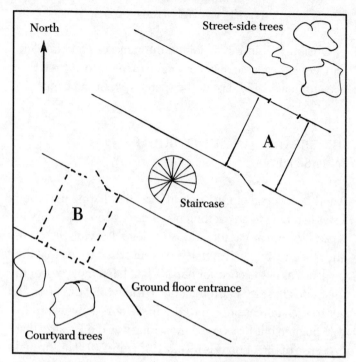

North   Street-side trees

A

Staircase

B

Ground floor entrance

Courtyard trees

FIGURE 51 A. *My room on the third floor with the window facing northeast and the street-side trees outside.*
FIGURE 51 B. *The illusion of my room on the third floor with the window facing the courtyard, from where I entered the building.*

Another likely factor is what one could call the special "spatial tiredness" that affects the spatial system as it encodes new cognitive maps and keeps track of directions and locations in a new environment. I had arrived in the area for the first time the day before.

The deterioration of the spatial system after seventy-six years of service must also have played a role. Directions are no longer sharply felt in old age. It is all more tentative, more elastic, more adaptable.

MISORIENTATION
IN ST. ANNE'S COLLEGE, OXFORD

I arrived at the college in April 1997 in the early afternoon to attend a conference, and got a room on the third floor in a student dormitory. There was no elevator but a spiral staircase, thus a good opportunity for me to derail my sense of direction as I climbed the stairs. In the room I felt that the window was facing the courtyard in the west (actually it was in the southwest, but it is a normal simplification to use the main cardinal directions when one makes a cognitive map of a new place). The next morning I was surprised to see the sun shine through the window. I realized that I must have gotten my sense of direction turned around, since the window had to be facing east (actually northeast) for me to see the sun through it at that time of day (Figure 51).

I recall that I had a strange feeling that something was wrong when I later climbed that spiral staircase and came to the corridor leading to my room.

The reversal of the direction frame happened—as usual, I would say—when I arrived tired after the journey and walked at least a kilometer from the bus station carrying all my luggage. When I entered the room and saw the trees outside the window, my spatial system must have jumped to the conclusion that they were the trees in the yard from which I had entered the building. In reality, however, the trees I saw were on the opposite side of the building and, as a result, my direction frame and the cognitive map of the room I encoded became reversed.

My hypothesis is that we have a tendency to reorient ourselves in a room in a building so that the windows face in the same direction as the entrance we used to get into the building. In other words, we tend to assume that the outside area seen from the room is the same as the one we were in just before entering the building.

There are two factors that influence the chances of making a mistake in orientation when entering a room on the opposite side of a building: (1) the landmarks seen from the window, and (2) the state of our direction frame.

If the area outside the window is very similar to the area we came from (in my case with the same kind of trees), mistakes can easily be made. Whereas if we enter the building from an open area and see the wall of another building through the window, a mistake is most unlikely.

Clearly the stronger the direction frame, the more unlikely it will be turned around by the appearance of landmarks outside the window. Thus the weakening of the direction frame by fatigue, travel, and deterioration with age will make mistakes more likely.

Also, the longer and more intricate the passage from the building entrance to the room, the more likely it is that the direction frame will be derailed. If, for example, I had not climbed the spiral staircase to get to my room, but instead had been on the ground floor just opposite the entrance, I am sure the reversal would not have happened.

REVERSAL
IN EL CENTRO REST AREA

Around nine P.M., after a four-hour drive from Phoenix—a rather stressful experience, especially when heavy rain drastically reduced the visibility—my wife and I exited at a rest area along I-8 some five miles west of El Centro. She was driving and had trouble finding the restrooms. Finally she noticed that we had passed them, so she made a U-turn and parked, facing east. When I came back from the restrooms, I was surprised to feel that we were on the south side of the freeway (we were actually on the north) and when my wife drove away after making another U-turn, I felt that we were going back east towards Phoenix.

When writing down my account of the event the day after, I could still recall my strange cognitive map of the rest area. It consisted of two unconnected parts—one, the exit from the freeway to the rest area on the north side and the first U-turn; the other, the parking of the car on the south side facing west, and afterwards the departure to the east following the second U-turn (Figure 52).

Since we had stopped in the rest area on the south side

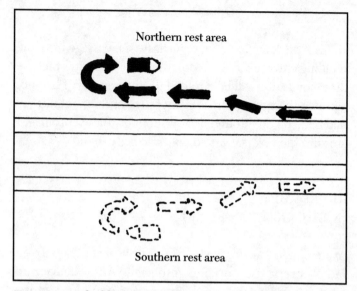

FIGURE 52. *The black arrows show us coming from the east and exiting to the rest area on the north side of the freeway.*

The white arrows show my illusion of our car being parked in the rest area on the south side of the freeway and getting back on the freeway in the eastbound lanes.

of the freeway two days earlier when driving east, I may have used the cognitive map encoded at that time in spite of being now on the north side, which reversed my sense of direction. After all, these rest areas, despite being on opposite sides of the freeway, are very much alike. My age and the dark, moonless night may also have contributed to the reversal.

The really interesting part, however, is what happened as we drove away. At first I felt we were going east, but by a mental effort I could "turn myself around" (i.e., turn my direction frame 180°) so that I felt, correctly, that we were going west. Oddly enough, I was able to make another "spatial U-turn" and thus get back into feeling, erroneously, that we were going east. It was as if I could alter at will my feeling of which direction we were going. After going back and forth a few times between the reversed and correct orientations, I finally settled down in the correct one.

If one compares this experience with the ones in Cologne and Paris, when I was totally at the mercy of my spatial system, it is obvious that by my mid-seventies, my spatial system had much deteriorated. My weakened direction frame easily gets turned around, but luckily this weakening also makes it easy to turn it right again. Maybe that is what those with weak spatial systems experience all their lives.

There is, however, an important difference. In Cologne and Paris, the violent "spatial vertigo" experienced when entering a reversed area was driven by actually seeing the landmarks encoded in the reversed cognitive map. In the car at night, there were no landmarks to be seen. Thus, I could easily picture the road as going in either direction. Be that as it may, I am still convinced that this rest area reversal could not have happened in my heyday.

REVERSAL ALONG
A DESERT TRAIL

On April 27, 2000, I was hiking in the McCain Valley along the Pepperwood Trail southwards towards Cottonwood campground. It was 10:15 A.M.—the sun was high enough to be out of sight for me wearing my sunhat—and I was tired after hiking with only a few breaks since eight A.M. It was hot, some 80 degrees in the shade, and there was no shade on the trail.

I was looking for a shady place to rest, and finally found a shrub approximately 5 meters away from the trail casting a large shadow within which I sat down. I got out my topographic map to see where I was and estimate how long it would take to get to the campground. I also looked intently for a while at a tiny flower, trying in vain to determine its species.

After about ten minutes I got up and walked back to the trail. When I reached it, I felt confused, uncertain which way to go, but finally turned right. After perhaps 50 meters I realized to my surprise and dismay that I was going in the wrong direction, back the way I had come.

Why this confusion when getting back to the trail after resting? Well, why am I not always confused? The answer is that my cognitive map, kept oriented by my direction frame, shows me clearly which way to go. Therefore the system that keeps the cognitive map oriented must have been shut down at the moment I reached the trail. Without this system, all I could do was guess and hope for the best; and, as it turned out, my guess was wrong.

Such a turnaround had never happened to me before when following a trail in nature. This indicates that the deterioration of my spatial ability in old age is at least partly responsible for the mishap, added to my exhaustion after hiking for over two hours in the hot sun. This tiredness, too, was enhanced by a lack of physical and mental stamina incurred in old age.

What happened in this case was not a slow drifting of the direction frame, such as occurs when we walk in circles when lost, but a temporary shutdown, so that when starting up again, the direction frame could settle in any position. When constrained only by a linear landmark, like a trail, it becomes either oriented correctly or is turned around. If there had been an eye-catching landmark visible when I returned to the trail, like a large boulder with a strange shape, it would have locked in my direction frame, and I would never have taken off in the wrong direction.

If this had not actually happened to me, I would not have imagined I could ever make such a mistake. Apparently age has rendered my sense of direction no longer reliable. I need the support of prominent landmarks to function properly.

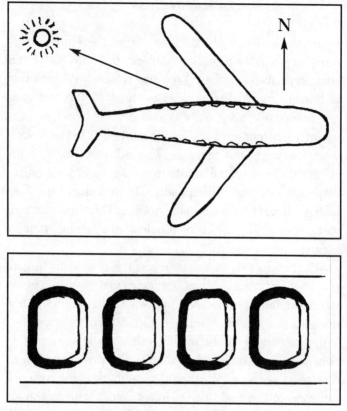

FIGURE 53, TOP. *The airplane flying eastwards with the evening sun low over the horizon in the west-northwest.*

FIGURE 53, BOTTOM. *The right side of the window frames illuminated by the sun shows that this must be the north side of the cabin. I still felt, however, that those windows were towards the south.*

FEELING A PLANE
FLYING BACK AND FORTH

I was flying one day in a plane from Chicago to White Plains, New York, thus on an easterly course. The plane had five seats in each row, two to the left and three to the right of the aisle. I was sitting in the aisle seat on the right, thus approximately in the middle of the row, and could therefore see just as well to the left as to the right through the windows. We were flying over a dense cloud cover, which hid the ground completely. After a while I noticed that when looking out of the windows to the left, I felt that we were flying westwards back to Chicago, whereas when looking out of the windows to the right, I felt correctly that we were flying eastwards. Thus, simply by moving my gaze from the windows on the right side to those on the left side, I could reverse my orientation, and by returning my gaze to the right side I could correct my orientation again. I found this extraordinary so I repeated the trial several times, always with the same results.

In this plane my spatial system acted as if it had no reliable landmarks to go by. My *knowledge* that the plane was flying eastwards clearly did *not* impress my spatial system. There was however one reliable "landmark" that ought to have prevented the reversal when I looked at the windows to the left, the north side of the plane. Since we were flying above the clouds, the sun near setting was shining from the west-northwest, illuminating the right side of the window frames, which therefore had to be the east side (Figure 53).

Apparently this fact was too cerebral to lock in my direction frame correctly. Instead, my spatial system went by the intuitive windows-in-the-south cue, which caused the reversal when I looked towards the left (north) side of the plane, and prevented it when I looked towards the right (south) side.

I find it rather astonishing that my direction frame is now so free to rotate that it can turn 180° back and forth without any apparent resistance. Does that mean that it is so unreliable that it could no longer keep me on course if I tried to rely on it, rather than my compass, when hiking cross-country? Or would it still function properly in a natural setting when, instead of being passively transported, I actively walked towards a goal?

It would be interesting to repeat the experiment I made in the desert when I managed to follow a surprisingly straight course with my sunhat pulled down to reduce my field of vision. Would I still be in the same class as Victor Cornetz's Saharan guide, whose "inner compass" never failed, or have I deteriorated to the level of Cornetz himself, who quickly lost his bearings after making detours around shrubs that were in the way?

CHAPTER 41

SPATIAL MEMORY SLIPS

CAUSING REVERSALS

In this chapter I look at three reports of reversals of orientation that indicate that they can be caused by a change of location or direction that slips out of the spatial memory.

Natalie Angier was walking down Lexington Avenue in Manhattan, and stopped to admire the St. Jean Baptiste church at the corner of Seventy-sixth Street. About ten blocks farther down the avenue, she crossed to the opposite side to buy juice in a store. She then continued walking until another church came into view, strangely similar to the first one. She writes:

How can it be, I ask myself, that there are two such similar and ornate churches on Lexington Avenue, and that I, a lover of architecture and longtime New Yorker, hadn't realized it? I run up to the entrance to explore this second interior, but stop short in horror when I read the sign: St. Jean Baptiste. Holy déjà vu all over again! With considerable self-disgust, I reconstruct what happened. The simple act of crossing Lexington Avenue for refreshments had completely flipped around my sense of left, right, north and south, with the result that, after leaving the grocery, I had mindlessly retraced my steps back uptown.[69]

To make the discussion easier, let us assume she was on the east side of the avenue when walking south (downtown) from the church. She would then have crossed over to the west side to shop. When she left the grocery, she turned left to continue downtown, which shows that at that moment she must have felt she was back on the east side. The reason for the reversal of her sense of direction was therefore that the act of crossing the street had slipped out of her spatial memory. Afterwards she walked uptown on the west side of the avenue, feeling she was going downtown.

Her failure to recognize the church the second time around is easy to explain. She simply didn't expect it to be there, for in her mind it was now located about twenty blocks farther south and on the other side of the avenue. In addition it was turned around 180°.

To be precise, it was not crossing the avenue, per se, that caused her sense of direction to turn around, it was her spatial system forgetting it. Nor was there anything wrong with her sense of left and right; it was her feeling that north was south that made her turn left instead of right when leaving the grocery.

The second report is by Jacques Passy, a chemist in Paris.

He was on his way to Rue de Rennes from the west, but crossed it without noticing it and continued until he realized he must have gone too far. He therefore turned around and walked back towards the west until he reached Rue de

Rennes. There he had a nasty experience. He could not understand how he could see Gare Montparnasse (at the south end of Rue de Rennes) on the left, when he felt it ought to be on the right. He gives a very good description of how he experienced this:

> My confusion was extreme, and I found myself for some time in a state of complete disorientation, and unable to comprehend the situation. This state did not at all resemble that which is produced in an unknown place—in a forest, for example, where one has gone astray and does not know where he is. Here you know very well where you are. You have a very clear sense of direction and you know perfectly where things ought to be; only this direction and this place are just the opposite of their real position. . . . When you recognize the objects you realize you are under an illusion, but the illusion does not disappear.[70]

Obviously, since he expected to see Gare Montparnasse to the right he must have felt he had reached Rue de Rennes from the west, his original approach. His crossing that street and later turning back towards it must therefore have slipped out of his spatial memory.

Another experience of reversal concerns that of my friend Louis, who had driven from Texas to Los Angeles over three days, and was going north on Interstate 405 when he exited at Roscoe Boulevard to buy gasoline, about one hour after nightfall. The gas station was on the west side of the freeway so he drove through the underpass to get there. Going back to the freeway afterwards, he turned right into the southbound on-ramp and started driving south on 405. Only when he saw the name of the next exit and remembered that he had already passed that place, did he realize that he was going south. After turning around he knew he was going in the right direction but he still felt he was going in the wrong direction for quite some time.

In this case also the cause for the reversal of the sense of direction was a slippage in spatial memory. It obliterated the drive through the underpass and instead located him on the east side of the freeway. He therefore felt he had to turn right onto the on-ramp to get back into the north-bound lanes. The fact that it was dark, and that he was tired after driving all day, must have made it easier for this to happen.

One discerns a general pattern in these reports. It starts with an established routine: proceeding for some time in a certain direction, unthinkingly, like a robot. Then the routine is broken. Passy turned around and started walking in the opposite direction, the other two stopped and crossed over to the other side of the freeway or avenue.

After the break, Passy slipped back unconsciously into the previous routine; in spite of having turned, he felt he was walking in the same direction as before. Louis slipped back in his mind to the east side of the freeway, where he had been driving all day, and accordingly took the wrong on-ramp. Angier slipped back in her mind to the east side of the avenue, where she had been walking earlier, and there-fore turned the wrong way when she left the grocery.

One could say that the attraction of the well-established

routine is such that the spatial system slips back into it directly, without taking the changes during the break into account, which leads to the reversals. These are not random mistakes; there is method in them; the slippage is the root cause.

CHAPTER 42

THE ROLE OF GESTALT

IN MISORIENTATIONS

I got the idea for this chapter after reading a paper by D.C.D. Pocock, a British geographer who studied the cognitive maps of residents and visitors by making them draw map sketches of the city of Durham from memory. He found that they had a tendency to straighten roads and make intersections at right angles, and attributes this to our "inherent tendency to 'betterment,' expounded in the gestalt school of psychology, and applicable to all forms of perceptual organizing."[71] They were thus following the gestalt law of good shape.

It struck me that applying this gestalt law of good shape could throw light on some of the misorientations I had studied. It would explain why we encode a winding road as straight, which led to the misorientations at Buckman Springs. Peterson's misorientation in Minneapolis can perhaps also be seen as caused by a good shape cognitive map of the railway going straight into the station, when in reality it turned 90°. The same goes for the misorientation that happened during the drive at night from Boston to Franklin when the roads were made straight and the intersections at right angles in the cognitive map encoded. I had better stress here that all of this takes place on the unconscious level.

The hiker who got his sense of direction turned around when he tried to cross a ridge near Siegen was following a good shape cognitive map of a ridge sloping evenly on both sides, thus to be crossed by going straight up to the crest and straight down on the other side. The lateral spur from the ridge that he ran into, and that was not on his good shape map, then led him astray, so that he came back down on the same side of the ridge.

But in some cases the gestalt law of good shape does not explain the misorientations. What is needed is a gestalt law of good orientation. I am thinking of a strange experience I had when reading in my notebook about a hike I had done three years earlier in the Cactus Garden in the Southern California desert. I had written that 'I started hiking towards the east, but I thought I must have mistaken the direction, for, as I recalled, I had started out towards the south. However, after a while it struck me that the note was indeed correct; it was the cognitive map I was using when reminiscing that was turned so that north was west and south east. That explained the mistake. But why did the cognitive map turn? My answer: it followed a gestalt law of good orientation. The valley in the desert area I had in mind runs east–west with "up the valley" towards the west, whereas the good orientation of a valley would be with "up the valley" towards the north. This might be linked to our habit of saying "up north" and "down south."

I remember a lady who told me how in her youth she

often went with her parents from Los Angeles "down into the desert." As the road started going down the valley into the desert, she felt they were going south, when in fact they were going east. In her mind she was thus following the gestalt law of good orientation stating that "down the valley" equaled "towards the south."

A friend of mine teaching geography in a high school told me that her students had trouble with the lay of the land in Egypt, the problem being that the Nile river flows towards the north, which is contrary to the gestalt law of orientation. This complicates matters: for example, where is Upper Egypt? Is it up north, or is it upstream towards the south? Very confusing! Not like the Mississippi River flowing in the proper direction with upstream up north and downstream down south. Her students had no problem with "Ol' Man River," since it obeyed the law of orientation.

The Laps in Northern Sweden are a case in point. They herd their reindeer along the river valleys from the lowland up to the mountains in the spring and down again to the lowland in the fall. The paramount directions in their cognitive compass rose are not north and south, but *davvi* = up the valley, which here is northwest, and *lulli* = down the valley, which here is southeast. This does not, of course, prevent them from using the sun for orientation; however, the directions they get from the sun are not thought of in terms of north and south, but in terms of *davvi* and *lulli*.

Going from the wilderness to the metropolis we encounter the same gestalt law of good orientation. As all New Yorkers know, the avenues from Lower to Upper Manhattan all run north–south, thus in a good direction, except on topographic maps, where they run towards the north-northeast. But who looks at topographic maps when in Manhattan? Even the street maps show the avenues running north–south, thus rejecting reality and obeying the law of good orientation.

The discomfort of a lady in Santa Barbara, where the coastline runs east–west so that the sun rises over the ocean in winter, could be seen as caused by the coast disobeying the law of good orientation. This law dictates that a west coast should run north–south.

My misorientation in the Pasadena auditorium, where I felt that the windows were in the south wall when in reality they were in the north wall, indicates that there is a gestalt law of good orientation maintaining that windows should be in the south wall of a room. Such a law would also give a beautiful explanation to my airplane experience, where I correctly felt the plane was flying east when I looked to the windows on the right, the south side of the cabin, and erroneously felt it was flying west when I looked to the windows on the left, the north side.

While the gestalt laws of shape are universal, this is not always the case with the laws of orientation that I have proposed. Some of them must instead be seen as local ordinances holding sway in limited areas. They would develop and serve well when individuals grow up and remain in an area. However, when they go somewhere else they might experience misorientation. A good example: people growing up on the East Coast develop a feeling that the ocean is always in the east, which can lead to reversals of orientation if they move to the West Coast.

In the southern hemisphere the sun is in the north at noon. One would therefore expect the preferred location for windows in a room to be in the north wall. Similarly, in the northern hemisphere a slope facing south is more attractive since it gets more light and warmth.

It could be that when our inner sense of direction deteriorates in old age, we become prone to turning unconsciously to the gestalt laws of good orientation for what one might call default directions.

CHAPTER 43

DO HUMANS HAVE

A MAGNETIC SENSE?

Whenever one contemplates our way-finding system, there is one question that sooner or later emerges and demands an answer: What do we go by to keep on course hiking cross-country when there is seemingly nothing to go by? For it is difficult to imagine that our system would be so incomplete that it would stop functioning when there were no outer cues for direction available—because when we get our directions wrong, the whole system inevitably fails.

I have described in this book what can go wrong with our sense of direction: the misorientations in which we feel, in a certain area, that north is where we know it is not; and the more dramatic situation of total reversal, in which we feel that north is south. Then there is the slow turning of our sense of direction when walking in an environment with limited visibility, like a forest, which misleads us into walking in a circle while feeling all the time that we are following a straight course.

I might have created the impression that these are fairly frequent mishaps, but the opposite is true. These are rare events. The remarkable thing is not that they happen, but that they happen so seldom. And when they do happen, there is generally an *inner* cause involved, like fatigue or jet lag.

All of which leads us to assume that there is an *inner* cue for direction that prevents those mishaps, that keeps us oriented when there are no outer cues to go by. For example, it could well be the case that we always have a tendency to deviate when trying to keep on course through a forest, but that this deviation is detected and corrected by an inner direction cue.

It could be that these misorientations and deviations occur when outer cues are lacking *and* when the inner direction cue is too weakened, for example by tiredness, to prevent it.

As to the nature of this inner cue for direction, the most likely candidate is a magnetic sense of some kind—hidden, one would imagine, at the very foundation of our way-finding system, far too deep to be reached by introspection; a kind of default direction cue that kicks in only when all outer cues are missing. I had better put in the usual caveat here, that it might not develop in modern urban subjects, and that it most likely reaches its full potential only in "primitive" people, those who really need it.

My second argument for humans having a magnetic sense is of a Darwinian nature: *sum ergo teneo*, I am therefore I have (it). For if my ancestors eking out a living in the Swedish forest had not had a magnetic sense that would give them an intuitive notion of which way to go in the middle of the forest when the sky clouded over, they would not have made it. They would have died out and I would not be around.

My third argument is this: It has been shown that several primitive species, e.g., mollusks, crustaceans, insects, frogs, newts, and fishes, have a magnetic sense. Hence if we go far enough back in our evolution, we are likely to find a primitive ancestral species that had a magnetic sense. This magnetic sense would not have disappeared as we evolved, unless there was a long enough period when we did not need it. I find it most unlikely that such a period existed. It is therefore highly probable that we still have a magnetic sense to fall back on when there are no outer cues for direction.

A question that naturally comes to mind here is: What kind of experiments would be likely to demonstrate that we have a magnetic sense? I know it is the normal practice in psychology to recruit subjects randomly "from the street," but since most members of civilized societies can manage quite well with even a very mediocre sense of direction, experiments with such subjects are likely to give inconclusive results. The positive results of a small minority of the subjects are likely to drown in the negative results of the majority in the statistical treatment of the data. This could have been what happened in Baker's experiments[72] where his conclusion, based on statistics, that humans have a magnetic sense, was hotly contested by other researchers who argued that his statistics were flawed and did not prove anything.

What must be done is therefore to screen the subjects, before the experiments, to eliminate those with a mediocre sense of direction. The ideal proving ground would be a plain with a fairly dense forest, where the subject would be asked to keep going in a certain bearing and the deviation after they had proceeded a certain distance would be measured. Where such an area is difficult to find, one could instead make experiments on an open plain at night, with the subjects using a flashlight for guidance. As my own experiences have shown, now that my sense of direction has deteriorated in old age, this can quickly lead to a large deviation. A route length of, say, 50 meters would probably be enough to sort out the incompetent. I would not recommend the use of blindfolds for these experiments. The method is much too unnatural and stressful. The subjects need to feel perfectly at ease to function optimally.

The subjects with a good sense of direction would then be given helmets: half of them with magnetic bars strong enough to distort the earth's magnetic field and the other half with nonmagnetic bars, and the experiment would be repeated. If those with magnetic bars in the helmet then deviated significantly more than without magnets, it would be a strong indication that they were relying on a magnetic sense in their way-finding.

If we have a magnetic sense, it would most likely be installed, to use computer language, before the systems based on outer cues, like the sun compass. There would then be a window in the development of the child when it had mostly the magnetic sense to rely on. If this assumption is correct, this would obviously be the ideal time for experiments to determine if humans have a magnetic sense. One could for example put a toddler in its mother's lap on a swivel chair moved back and forth at angles between, say, 30° and 60°, while the mother covered the child's eyes with her hands. The sense of direction detector could be a fixed ring of boxes surrounding the chair, all containing the same toy, one of which the child had seen hidden before the experiment.

Between the swiveling sessions, the child would be encouraged to reach for the box where it felt the toy would be, take it out and play with it. If the toddler had a good sense of direction, it would reach for the same box after each session. By diverting the earth's magnetic field with Helmholtz coils,[73] one could then determine if the toddler used a magnetic sense.

I predict that one day experiments will have been made that prove beyond reasonable doubt that we do have a magnetic sense.

The resistance against admitting that humans might have a magnetic sense could in part be due to the reluctance of those with a mediocre modern-urban sense of direction to believe that others might be better endowed. I can understand their view, now that my own sense of direction has deteriorated in old age, but I cannot share it, because of my memories of how well it functioned in my heyday.

CHAPTER 44

SUMMING UP

AND LOOKING AHEAD

Having arrived at the end of this book, it is time to sum up what we have discussed, and to look ahead towards what remains to be explored.

First I have a confession to make. This is not my forte. Summaries are for *orderly* people, those who love to take old pieces of knowledge and shuffle them around until a picture emerges. Whereas I much prefer to scurry hither and thither aimlessly, and, when I am lucky, stumble on new insights. All I can do is pick up the pieces and polish them off a little before I present them.

The main ingredients of our way-finding system are the cognitive map, the direction frame—or "sense of direction" —and the dead reckoning system—or "sense of location."

The cognitive map is encoded automatically when we explore a new area. For this map to be oriented, the direction frame has to stay oriented. For the landmarks to become encoded in the proper location, the dead reckoning system has to function properly.

One can see the cognitive map as the readout from a database made up of the landmarks with which we are familiar—their shape, location, and orientation. This read-out is seen from where we are and in the direction we are looking. It is therefore not seen from above, as on a topographic map, but from the side, from approximately eye level. It therefore shows landmarks the way they look when they appear ahead as we proceed.

This readout from our "spatial computer" is extremely user-friendly. We don't need to be computer nerds! A child can use it, animals use it.

A cognitive map is very robust, difficult to change. While the aspect of a landmark in the map can change, such as a tree in winter after its leaves have fallen off, the location and orientation of a landmark cannot readily be changed.

As we become increasingly familiar with an area or a route, for example one close to home, the role of landmarks as cues for location and direction diminishes. Instead we rely more on dead reckoning to get a feel for where we are. We therefore start forgetting landmarks in such an area, since we no longer need them.

However, those with an undeveloped "modern-urban" spatial ability need the constant reassurance of familiar landmarks to find their way (unless they use a map and compass). Primitive people can fall back on their well-developed basic spatial system and go for long distances without relying on landmarks. For them, landmarks only serve as a confirmation that this spatial system is working properly.

There are two kinds of landmarks: those we know so well that we can describe them in advance, and those we recognize only when we see them. I call them active and passive landmarks. When we follow a familiar route, we are able to visualize the active landmarks before they appear. The passive landmarks we do not anticipate seeing, but we remember that we have seen them before when they come into view.

Normally we actually see only a small area in the real world, while the cognitive map shows us all the areas we are familiar with. In a way, we therefore function much more in our cognitive map than in the real world. It is not only the picture we make of our environment, it is the link between our personality and our environment, our foothold in it. One could even say that the cognitive map is the part of our personality that links us with our environment. The panic I felt when my dead reckoning system slipped so that my cognitive map no longer linked me to my familiar environment showed me forcefully—I would even say brutally—how vital this linkage is. Without it we would feel utterly lost, even in our own home!

When we decide to go to a place, we first assess its direction and distance from where we are. We then look at a detailed cognitive map of the place by imagining that we are there walking around and "seeing" the landmarks. Finally we can look at the cognitive route map to follow in order to get there. This map is a series of views we will see along the road, both the interesting and beautiful ones that reassure us that we are on track, and the really important ones showing crossroads and forks, with the road we have to take clearly marked. Note that for the return we have a very different route map, since we now come from the opposite direction.

Sometimes, but luckily very rarely, things go wrong even

for the best of way-finders. They can end up with a reversed direction frame and then encode a cognitive map of the area they visit that is also reversed so that they *feel* that north is south, in spite of *knowing* very well that this is wrong. They then find that this map stays misoriented. Every time they return to the area their direction frame will spin around 180°, inducing a kind of "spatial vertigo."

It can also happen that their direction frame comes loose, for example when hiking in a forest. At such a time, it does not turn randomly back and forth but drifts steadily in one direction. Misled by the direction frame the hiker will then veer steadily to one side and end up walking in a circle.

Our way-finding ability increases as we grow up, and probably reaches a plateau in middle age. Since my own way-finding ability deteriorated considerably after seventy years of age, this could be a general trend, albeit with individual differences in onset and severity of the decline.

Human way-finding ability is a tricky subject to study since it varies greatly; some people "never" get lost and others "always" get lost. There are at least two reasons for this:

1. In the Paleolithic, the selection pressure was on the *group*, not on the individual. The lone hunter with his feeble weapons was more likely to end up as prey than remain a predator. There was safety in numbers, and only the hunting band could survive. It was therefore enough if a couple of hunters in the group could find their way home to the camp; the others could just follow.

2. Our urban civilization has made our primitive way-finding ability superfluous. Few of us ever go cross-country. Even in wilderness parks we obediently follow the rules and stay on trails. Since the brain is opportunistic, our spatial ability does not develop to anywhere near its full potential.

LOOKING AHEAD

This is not all I have to say about way-finding. It is all that fits in one book of normal size. Having summed this one up, it is time to look forward to the next one.

First there is the fascinating subject of primitive ocean navigation that demands to be looked at. How did the Norsemen cross the Atlantic to the American continent long before Columbus? And how did the Polynesians and Micronesians manage to reach and colonize every habitable island in the Pacific, even the very tiny ones? This demanded precise navigation, indeed, and they did it with Stone Age technology.

Then there is way-finding with map and compass in the sport of orienteering, seemingly a purely cerebral affair, but one in which elite orienteers are able to transform their topographic map quickly into a cognitive one, and then rely on their direction frame when running so that they do not lose time by having to stop frequently and check with the compass that they are on course.

Finally, there are the many experiments that have been made on infants and toddlers and on animals, especially laboratory rats, that show that we start out in life with the basic "animal" navigation system, relying on the direction frame and dead reckoning, and later develop a more sophisticated adult system in which landmarks play a greater role.

All these topics deserve to be explored, and as I now thank you, my dear reader, for your attention, and close this book, it is therefore not by saying a final *adieu*, but rather a hopeful *au revoir.*

NOTES

1. Jonsson 1993.
2. Gatty 1958, 155.
3. Jonsson 1993.
4. Binet 1894, 344. My explanation in brackets.
5. Bonnier 1900, 76, my translation.
6. Cornetz 1909b, 303, my translation.
7. Cornetz 1929, 362, my translation.
8. Cornetz 1910, 154–155, my translation.
9. Gallistel 1990, 97.
10. Cornetz 1909a, 61–62, my translation.
11. Unpublished letter from Cornetz (1931), cited by Jaccard 1932, 269, my translation.
12. Van Gennep 1909, 300, my translation.
13. Dodge 1877, 52–53.
14. Ibid., 66.
15. Irwin 1985, 183, quoted with permission from the publisher.
16. Ibid., 181, quoted with permission from the publisher.
17. Van Gennep 1911, 48, my translation.
18. Cornetz 1909a, 65, my translation.
19. Passini 1984, 27.
20. Anderson 1983, 127–128.
21. Frere 1870, 185.
22. Ibid., 186.
23. I am indebted to Victor Cornetz for finding this reassuring declaration that in spite of all progress in science, there will always be mysteries left to solve.
24. Wrangell 1841, 140–141.
25. Darwin 1873, 418.

26. Bastian 1880, 215.
27. Viguier 1882, 16, my translation.
28. Jaccard 1932, 145, my translation.
29. Lewis 1976, 262, quoted with permission from the publisher.
30. Ibid., 262.
31. Irwin 1985, 179–180, quoted with permission from the publisher.
32. Skinner 1942, 9.
33. Hyltén-Cavallius 1863–64, 278, my translation.
34. Ibid., 279–280.
35. Bonnier 1900, 86, my translation.
36. Catlin 1867, 96–98.
37. Dodge 1877, 48–49.
38. Howard and Templeton 1966, 258.
39. Schaeffer, 1928, 294.
40. Ibid., 299–300.
41. Darwin 1873, 418.
42. Forde 1873.
43. Viguier 1882, 19–20, my translation.
44. Ibid., 20.
45. Binet 1894, 338.
46. Ibid., 340–341.
47. Ibid., 341.
48. Ibid., 347.
49. Ibid., 338.
50. Ibid., 342.
51. Ibid., 343.
52. Ibid., 343.
53. Ibid., 345–346, my clarifications in brackets.
54. Ibid., 349.
55. Ibid., 350.
56. Peterson 1916, 229–230.
57. Ibid., 230–231.
58. Trowbridge 1913, 894.
59. Peterson 1916, 231.
60. Gatty 1958, 154.
61. Peterson 1916, 234–235.
62. Binet 1894, 345.
63. Passini 1984, 40.

64. I should of course give the reference here, but, alas, I was unable to find it. Instead I found the old Chinese animal symbols for the cardinal directions, which might have been on those banners: the dragon for the east, the bird for the south, the tiger for the west and the turtle for the north (Saussure 1928, 68). Incidentally, this is another indication of the paramount importance in antiquity of east, the mighty dragon, and the insignificance of north, the lowly turtle!

65. Baumgarten 1927, 115–116, my translation.

66. Kirschmann 1926, 245, my translation.

67. Ibid., 253.

68. Wiltschko and Wiltschko 1991, 444, my translation.

69. Angier 1999, F1.

70. Binet 1894, 340.

71. Pocock 1976, 502.

72. Baker 1981 and 1989.

73. Two identical, round, flat coils some distance apart with a common axis that give a very uniform magnetic field when connected in series to a direct current source. By arranging them with the axis horizontal and perpendicular to the earth's magnetic field and regulating the current, the desired deviation of the earth's magnetic field between the coils can be obtained.

BIBLIOGRAPHY

*(The chapters where the items have been referred to
are given in brackets.)*

Anderson, E. W. 1983. *Animals as Navigators*. London: Van Nostrand Reinhold Co. [Ch. 18]

Angier, N. 1999. Directionless? Scientists Offer Some Clues. *New York Times,* June 22, 1999, pp. F1–2. [Ch. 41]

Baker, R. R. 1981. *Human Navigation and the Sixth Sense*. London: Hodder & Stoughton. [Ch. 43]

Baker, R. R. 1989. *Human Navigation and Magnetoreception*. Manchester and New York: Manchester University Press. [Ch. 43]

Bastian, H. C. 1880. *The Brain as an Organ of the Mind*. London: Kegan Paul & Co. [Ch. 22]

Baumgarten, F. 1927. Die Orientierungstäuschungen. *Zeitschrift für Psychologie* 103:111–122. [Ch. 35]

Binet, A. 1894. Reverse Illusions of Orientation. *The Psychological Review* 1:337–350. [Ch. 5, 28, 29, 30, 31, 32, 34, 41]

Bonnier, P. 1900. *L'orientation*. Scientia. Série Biologique no. 9. Paris: Carre et Naud. [Ch. 7, 24]

Catlin, G. 1867. *Life among the Indians: A Book for Youth*. London: Sampson Low, Son, and Marston. [Ch. 25]

Cornetz, V. 1909a. Observations sur le sens de la direction chez l'homme. *Revue des Idées* 1909 Deuxième semestre:60–65. [Ch. 14, 17]

Cornetz, V. 1909b. Note complémentaire aux Observations sur le sens de la direction chez l'homme. *Revue des Idées* 1909 Deuxième semestre:302–307. [Ch. 7]

Cornetz, V. 1910. *Trajets des fourmis et retours au nid*. Institut général psychologie zoologique: mémoires 2. Paris: La Société. [Ch. 8]

Cornetz, V. 1929. Orientation, conservation de la direction, marche compensée, polarisation. *Journal de psychologie normale et pathologique* 26:354–409. [Ch. 7]

Darwin, C. 1873. Origin of Certain Instincts. *Nature* 7:417–418. [Ch. 22, 28]

Dodge, R. I. 1877. *The Hunting Grounds of the Great West*. London: Chatto & Windus. [Ch. 16, 25]

Forde, H. 1873. Sense of Direction. *Nature* 7:463–464. [Ch. 28]

Frere, Sir H. Bartle E. 1870. Notes on the Runn of Cutch and the Neighbouring Region. *Journal of the Royal Geographical Society* 40:181–207. [Ch. 19]

Gallistel, C. R. 1990. *The Organization of Learning*. Cambridge, MA: MIT Press. [Ch. 13]

Gatty, H. 1958. *Nature Is Your Guide*. New York: Dutton & Co. [Ch. 1, 34]

Howard, J. P. and W. B. Templeton. 1966. *Human Spatial Orientation*. London, New York, Wiley. [Ch. 27]

Hyltén-Cavallius, G. O. 1863–64. *Wärend och Wirdarne*. Första delen. Stockholm: P. A. Nordstedt & Söner. [Ch. 24]

Irwin, C. 1985. Inuit Navigation, Empirical Reasoning and Survival. *Journal of Navigation* 38:178–190. [Ch. 16, 23]

Jaccard, P. 1932. *Le sens de la direction et l'orientation lointaine chez l'homme*. Paris: Payot. [Ch. 14, 17, 22]

Jonsson, E. 1993. A simple model of our spatial system. Paper no. 6 in *RIN 93, Orientation and Navigation: Birds, Humans and Other Animals*. Oxford: The 1993 conference of the Royal Institute of Navigation.

Kirschmann, A. 1926. Über eine Orientierungstäuschung. *Zeitschrift für Psychologie* 100:244–253. [Ch. 36]

Lewis, D. 1976. Observations on Route Finding and Spatial Orientation Among the Aboriginal Peoples of the Western Desert Region of Central Australia. *Oceania* 46:249–282. [Ch. 23]

Passini, R. 1984. *Wayfinding in Architecture*. New York: Van Nostrand Reinhold Co. [Ch. 17, 34]

Peterson, J. 1916. Illusions of Direction Orientation. *The Journal of Philosophy, Psychology and Scientific Methods* 8:225–236. [Ch. 33, 34]

Pocock, D.C.D. 1976. Some Characteristics of Mental Maps, an

Empirical Study. *Transactions of the Institute of British Geographers* 1:493–512. [Ch. 42]

Saussure, L. de. 1928. *L'origine de la rose des vents et l'invention de la boussole*. In Ferrand, G. *Introduction à l'astronomie nautique arabe*. Paris: Librairie Orientaliste Paul Geuthner. [Ch. 34]

Schaeffer, A. A. 1928. Spiral Movement in Man. *Journal of Morphology and Physiology*. 45:293–398. [Ch. 27]

Skinner, W. H. 1942. The Old-time Maori. *N.Z. Surveyor: The Journal of the New Zealand Institute of Surveyors*. 18:Dec. 6–9. [Ch. 23]

Trowbridge, C. C. 1913. On Fundamental Methods of Orientation and "Imaginary Maps." *Science* 38:888–897. [Ch. 34]

Van Gennep, A. 1909. Du sens d'orientation chez l'Homme. *Revue des Idées*. Deuxième semestre:298–302. [Ch. 16]

Van Gennep, A. 1911. *Religions, Moeurs et Légendes*. Second Edition. Paris: Société du Mercure de France. [Ch. 17]

Viguier, C. 1882. Le sens de l'orientation et ses organes chez les animaux et chez l'homme. *Revue Philosophique* 14:1–36. [Ch. 22, 28]

Wiltschko, W. and R. Wiltschko. 1991. Der Magnetcompass als Komponente eines komplexen Richtungsorientierungssystems. *Zoologische Jahrbücher. Abteilung für allgemeine Zoologie und Physiologie der Tiere* 95:437–446. [Ch. 36]

Wrangell, F. P. 1841. *Narrative on an Expedition to the Polar Sea in the Years 1820, 1821, 1822, and 1823 Commanded by Lieutenant, now Admiral Ferdinand Wrangell of the Russian Imperial Navy*. New York: Harper and Brothers. [Ch. 22]

INDEX

ABOUT THE AUTHOR

Erik Jonsson was born in Dicka, a little village near Avesta in the forest of Sweden, in 1922. The forest was his backyard and at the same time his training ground for wayfinding. In 1969, Jonsson emigrated with his family to California, where he worked in a materials lab. During this time, Erik explored his love of botany by leading many field trips for the California Native Plant Society, especially in the desert. Since his retirement in 1987, he has been studying navigation on land and at sea, and has taken part in conferences in Oxford in 1993, 1997, and 2001 on animal and human navigation. He lives in San Diego.